Cambridge Tracts in Mathematics and Mathematical Physics

GENERAL EDITORS

J. G. LEATHEM, M.A.
E. T. WHITTAKER, M.A., F.R.S.

No. 1

Volume and Surface Integrals
used in Physics

VOLUME AND SURFACE INTEGRALS USED IN PHYSICS

by

J. G. LEATHEM, M.A.
Fellow and Bursar of St John's College

Second Edition
With two additional Sections

Cambridge:
at the University Press
1913

CAMBRIDGE
UNIVERSITY PRESS

32 Avenue of the Americas, New York NY 10013-2473, USA

Cambridge University Press is part of the University of Cambridge.

It furthers the University's mission by disseminating knowledge in the pursuit of education, learning and research at the highest international levels of excellence.

www.cambridge.org
Information on this title: www.cambridge.org/9781107493810

First edition 1905
Second edition 1913
First published 1913
Re-issued 2015

A catalogue record for this publication is available from the British Library

ISBN 978-1-107-49381-0 Paperback

PREFACE TO THE SECOND EDITION

THE present edition of this Tract differs from the first edition only by the inclusion of two additional Sections. One of these deals with Gauss's theorem of the surface integral of normal force in the Theory of Attractions. The other discusses some theorems in Hydrodynamics, and includes a short account of the theory of 'suction' between solid bodies moving in liquid.

The author's arrow notation for passage to a limit, since its publication in the first edition of this work in 1905, has been adopted by many writers on Pure Mathematics, and may be regarded as now well established. Its application has rightly been confined to continuous passages to limit, and there is evidently room for some corresponding symbol to indicate saltatory approach to a limit value. A dotted arrow might perhaps appropriately serve this purpose; it would present no difficulty to the printer, but it is just doubtful whether it would be convenient in manuscript work.

The author desires again to express his thanks to Dr T. J. I'A. Bromwich for help in the preparation of the first edition of this work, more particularly for valuable suggestions with reference to the discussion of tests of convergence in § 13 and to the restriction upon f' in the theorem of § 38.

J. G. L.

ST JOHN'S COLLEGE, CAMBRIDGE.
October, 1912.

CONTENTS

VOLUME AND SURFACE INTEGRALS USED IN PHYSICS

Introduction

1. The student of Electricity, and of the theory of attractions in general, is constantly meeting with and using volume integrals and surface integrals; such integrals are the theme of the present tract. It is proposed, in the first instance, to examine how far it is justifiable to represent by such integrals the potential and other physical quantities associated with a body which is supposed to be of molecular structure; and, in the second place, to give proofs of certain mathematical properties of these integrals which there is a temptation to assume though they are not by any means as obvious as the assumption of them would imply. Illustrations will be taken for the most part from the theory of the Newtonian potential, and from Electricity and Magnetism; and attention will be directed, not to all the peculiarities of integrals which can be imagined by a pure mathematician, but only to those difficulties which constantly present themselves in the usual physical applications.

I. On the validity of volume-integral expressions for the potential and the components of attraction of a body of discontinuous structure.

2. The generally accepted formulation of the Newtonian law of gravitation is that two *elements of mass*, m and m', at a distance r apart attract one another with a force $mm'r^{-2}$ in the line joining them*. This statement of the law may be regarded as a generalisation founded

* In order to shorten the formulae the constant of gravitation is omitted here and elsewhere; its presence or absence in no way affects the questions to be discussed.

on the observed motions of heavenly bodies, and its simplicity commends its adoption as the starting point of mathematical discussions of gravitation problems. But the fact must not be ignored that the statement is really lacking in precision; for in the first place the phrase 'element of mass' is somewhat vague (even when the term 'mass' is sufficiently understood), and must be taken to mean simply a very small body or portion of a body, so small, namely, that its linear dimensions are very small compared with r; and in the second place r itself is rather indefinite, meaning the distance between some point of the first element and some point of the second.

The principle of superposition, that is to say the assumption that the force exerted on one element of mass by two others is that obtained by compounding according to the parallelogram law the forces that would be exerted on it by each of the other elements alone, is part of the fundamental hypothesis of the Newtonian law; and the principle is commonly used to evaluate the attraction of a body which is not extremely small by compounding the attractions of the small component elements of mass of which it may be regarded as built up. In works which deal with the mathematics of the gravitation potential and attractions, the values of these quantities for a body of definite size are invariably obtained in the form of volume (or surface) integrals taken through the space occupied by the body, the element of mass being represented by the product of the element of volume and a function of position called the 'density.' But it ought to be clearly understood that this procedure virtually involves either using the 'element of mass' of the Newtonian law as an *element of integration*, and thereby attributing to it properties which are directly contrary to accepted views as to the constitution of matter, or else using the word 'density' in a special sense which is by no means simple or precise. For there is no limit to the fineness of the subdivision of a region into volume elements for purposes of integration, and the process must get endlessly near to a limit represented by vanishing of the volume elements; if this extreme subdivision cannot be applied equally to mass, there comes a stage in the process when a volume element becomes too small to contain a mass element, and so the average density in the element, mass divided by volume, ceases to have a meaning, and the mathematical passage to limit which constitutes the usual definition of the density at a point is now impossible.

If the body considered is of mathematically continuous structure, so that the portion of it occupying *any space however minute* has the gravitation property, then density is a term having a precise meaning; but if the distribution of the gravitation property through the space occupied by the body has not this mathematical continuity, we cannot attach any meaning to the volume integrals till we have first invented a suitable new meaning for the term 'density'; and the inevitable vagueness that will arise in the new definition will preserve in the final results that slight lack of precision present in the terms of statement of the gravitation law, which might at first sight appear to have dropped out of results represented by such precise mathematical expressions as volume integrals.

It is practically certain that no substance can be subdivided without limit into small portions each of which possesses the gravitation property. There must be a stage of subdivision beyond which the component portions cease to have the properties of larger portions of the substance, and we may speak of the smallest portion of a substance that has the gravitation property as a 'particle' for purposes of the present discussion. What the order of magnitude of a particle may be it is difficult to guess, but the kind of generalisation from large bodies to small bodies which led to the conception of an element of mass suggests the possibility that the process of subdivision without loss of the gravitation property might be continued till we arrive at the molecule of Chemistry or Gas Theory. There is no experimental evidence to prohibit, and possibly some to justify our carrying the generalisation so far (provided we set some limit to the smallness of the distance at which the attraction of two particles is supposed to obey the law of the inverse square), and the great simplicity of the law thus obtained makes it an interesting one to study.

If a body has not got mathematical continuity in the distribution of the gravitation property throughout its volume, and so is to be regarded as made up of particles, we may speak of it as having 'discontinuous' structure. And if it be supposed that the particles may be as small as molecules, we must form a mental picture of the structure in which there appears no trace of material continuity, the substance being represented by discrete molecules or systems occupying spaces with somewhat indefinite boundaries, separated by more or less empty regions which may be called intermolecular space; (if the molecules move it may be supposed that their motions do not affect the properties under consideration). The problem of finding the potential or attraction of such a body at any point, if formulated on a greatly magnified and coarser scale, would in some respects resemble the problem of evaluating the potential or attraction of a mass of sand or other granular matter. We want to see how volume integrals present themselves as approximate solutions of such problems.

It is unnecessary to dwell here upon the familiar definitions of the intensity of force at a point, or the potential at a point, due to a gravitating body. But, for future reference, we may emphasize the fact that, among the mathematical properties of the potential at a point outside the body, that which may be taken as the fundamental physical definition is the fact that its space-gradient at any point is vectorially equal to the intensity of force there. If a point is so situated that a physical definition of intensity of force there is impossible, this physical definition of potential breaks down, and we are at liberty to substitute some convenient purely mathematical definition for which it may be possible afterwards to find a physical interpretation.

The potential of a body of discontinuous structure at an external point P is the sum of the potentials at P due to the particles that compose the body, i.e. Σmr^{-1}, where m is the mass of a particle and r its distance from P. Here, in accordance with what has been said above, r is not precisely defined, and a corresponding lack of precision must be present in Σmr^{-1}. By assuming P to be not too close to any particle of the body we can ensure that each r shall always be great compared with the linear dimensions of the corresponding particle.

When we endeavour to compare the values of this expression for the potential at different points, we recognise that the sum of a finite but extremely great number of extremely small terms is a most troublesome function to work with, and so there naturally suggests itself the device of getting a probably very approximate equivalent function by replacing the sum by the limit to which it would tend if the number of terms could be increased indefinitely while each separate term decreased correspondingly; this process would give us the potential in the form of the definite integral $\int \rho r^{-1} d\tau$, where $d\tau$ is an element of volume and $\rho\, d\tau$ the corresponding mass.

But, as has been suggested already, the transition from a sum of terms to a definite integral would imply the possibility of *endless* subdivision of the material mass into elements each possessing the gravitation property, whereas it is practically certain that matter cannot be so endlessly subdivided. In fact the use of the definite integral form implies a regarding of matter as continuously extended through the space which it effectively occupies, and attributes to the density ρ at any point the value obtained by passing to a mathematical limit in the usual fashion, that is to say the limit of the quotient of mass by volume for a region surrounding the point as the dimensions of the region tend to zero. The molecular view, however, requires us to cease subdividing matter beyond a certain stage, and so prevents our ever arriving at the kind of limit which is known as an integral.

3. Nevertheless the potential at the point P of an assemblage of discrete particles in a finite region may be equal in value to a volume integral taken throughout the region if the integral be supposed to refer to a hypothetical continuous medium occupying the same region and having a suitably chosen density at each point. It is only necessary to choose the law of density properly, and to this end there suggests itself the device of taking for each point A some sort of average density, based on a consideration of all the masses within a very small but finite region surrounding A. The dimensions of this small region might be settled by convention, but we need only consider the order of magnitude of these dimensions.

The kind of smallness that we want in this connexion is what we may call *physical smallness*, as distinguished from mathematical smallness to which there is no limit. Physical smallness implies smallness which appears extreme to the human senses, but it must not be a smallness so extreme as to necessitate passage from molar physics to molecular physics; it must leave us at liberty, for example, to attribute to matter occupying a physically small space the properties of matter in bulk if these should be different from the properties of isolated molecules. In fact a physically small region, though extremely small, must still be large enough to contain a very great number of molecules. It is estimated that a gas, at normal temperature and pressure, has about 4×10^{19} molecules per cubic centimetre; thus a cube whose edge is 6×10^{-5} centimetres (roughly the wave length of sodium light) contains more than 8,000,000 molecules; if we regard a million as a large number, the wave length of sodium light is (for other than optical purposes) physically small, and it is known that very much greater lengths than this appear to our senses extremely small. We are therefore in a position to speak of lengths which, though extremely small, are very great compared with other physically small lengths.

Now it is not suggested that the gravitation property is a molar property of matter, not possessed by a single molecule, for we have adopted just the opposite hypothesis; and so it might be thought justifiable to make the region round A, used for calculating ρ, smaller than merely physically small. This point will be referred to again, but at present it suffices to remark that the ρ generally used in potential theory is a continuous function whose value is not subject to very rapid fluctuations as A moves from one position to another. To get such continuity and smoothness in the suggested average, and to avoid the difficulty that would arise if the number of molecules intersected by the boundary of the region so as to be neither obviously inside nor obviously outside were not very small in comparison with the total number inside, we must take account of a large number of molecules at a time, and so we have to assume that the region surrounding the point A and used in getting a value for ρ is only physically small.

As regards the system of averaging, it is clear that in the case under discussion we get the best agreement between the potential of the actual and that of the hypothetical system if we give to each molecule an importance proportional to the product of its mass and the reciprocal of its distance from P; but it would be unfortunate to be obliged to average in this fashion, as we should thereby get a value of ρ at A which would not be independent of the position of P. And we should get quite a different law of density if we were dealing with some other integral, say an attraction integral, instead of that representing potential.

So long, however, as P is at a distance from A great compared with the linear dimensions of the physically small region used for the purpose of averaging, the values of r^{-1} for the molecules in this region are very nearly equal, and so there is very little error if in taking the average we give to each molecule an importance simply proportional to its mass. We thus get for the density ρ of the hypothetical continuous medium the quotient of mass by volume for the physically small region considered. This value of ρ has the advantage that it is independent of the position of P, and of the particular physical quantity whose integral expression is being investigated. But its great advantage, and the real reason why we adopt it, is that it is that density of a substance which we actually arrive at by practical methods of measurement; for ordinary laboratory measurings and weighings are applied to portions of a substance which are far from the limits of physical smallness, and so give us, not the sizes and masses of individual molecules, but only the total mass and the total space effectively occupied.

4. So far we have considered only the case in which the point P at which the potential (or other function of position) is to be estimated is at a distance from the nearest portion of the gravitating mass great compared with the linear dimensions of what we have called a physically small region. Such a distance, which, as we have seen, may be extremely small compared with the smallest distance we can measure directly, would seem to mark the limit of nearness of P to the gravitating body if the integral taken for the hypothetical continuous medium is to serve as equivalent to the true potential. But further consideration may enable us to push this limit still closer to the body. For the inaccuracies whose importance is magnified by decreasing distance do not, for a given position of P, occur in the case of each molecule of the body; they arise only in connexion with the molecules that are near P. Now such molecules, though perhaps absolutely numerous, are generally few in comparison with the remaining molecules of the body, and it is possible that their numerical inferiority may prevail over the advantage of their position in such a way as to render the total inaccuracy corresponding to them a negligibly small fraction of the whole potential.

Instead of the function r^{-1} which occurs in expressions for the potential, let us consider some other function f of position relative to P, which tends, as r becomes smaller, to become infinite of the same order as $r^{-\mu}$, so that, for small values of r, f is of the form $kr^{-\mu}$ where k is a finite function of relative angular position. Taking, in the first instance, a single physically small element of matter, say of volume ϵ^3, it is clear that the difference between the sum Σmf and the integral $\int \rho f d\tau$ through the element will in general be of the same order of magnitude as either quantity separately so long as r for all points of the element is of the same order of magnitude as ϵ, but that the difference will diminish to a quantity smaller in a ratio comparable with ϵr^{-1} when r becomes great compared with ϵ. Hence, for purposes of estimating order of magnitude, it is fair to represent the difference between Σmf and $\int \rho f d\tau$ by the expression $\int A \epsilon r^{-1} \rho f d\tau$ where A is a finite number.

To include all elements of the body near to P, we suppose the least value of r for a molecule near to P to be η, and take the integral representing inaccuracy through all space between the concentric spheres $r = \eta$ and $r = a$, where a is large compared with ϵ. If ρ' is the greatest value of ρ in this space, the order of magnitude of the inaccuracy is the same as or less than that of

$$\rho' k' A \int_\eta^a \epsilon r^{-1} r^{-\mu} 4\pi r^2 dr,$$

where k' is a finite constant replacing k, and $4\pi r^2 dr$, the volume between the spheres of radii r and $r+dr$, takes the place of $d\tau$. This is equal to

$$4\pi\rho' k' A\epsilon\{a^{2-\mu}-\eta^{2-\mu}\}/(2-\mu),$$

of which, when η is of the same order as ϵ, the first term is small of order ϵ and therefore negligible always, while the second term is small of the same or higher order provided $\mu<2$; the second term would be small, but not of so high an order of smallness, if μ were between 2 and 3. The case of $\mu=2$ would turn on the order of magnitude of $\epsilon\log\eta$ or $\epsilon\log\epsilon$, which is very small though not as small as ϵ. Sometimes the special form of the function k, taken in connexion with probable symmetry in the average distribution of molecules round P, increases still further the order of smallness.

It follows, therefore, that if $\mu<2$ the inaccuracy is certainly as negligible as ϵb^{-1}, and that if $\mu<2\frac{1}{2}$ the inaccuracy is certainly as negligible as $\sqrt{\epsilon b^{-1}}$, where b is some length which is physically not small, e.g. a centimetre. The case of $\mu=1$ corresponds to potential; for attraction components $\mu=2$.

5. Thus it appears that the representation of potentials and attractions by means of integrals extended through the hypothetical continuous medium which replaces the actual gravitating body is valid without sensible error not only for points well outside the body, but also for points whose distance from the nearest portion of the body is small of the order of the physically small length ϵ. This includes the case when P is so close to the apparent outer surface of the body as to be sensibly just not in contact with it, and also the case when P is in a small but not imperceptibly small cavity cut in the body, that is a cavity of such a size that the piece excavated would have the properties of matter in bulk rather than the properties of a few molecules.

As might be expected, any attempt to justify the use of the same integral expressions for the potential and attractions at a point P which is at a distance from the nearest molecule of a higher order of smallness than ϵ, results in failure. For now a simple molecule contributes to the potential (for example) a term mr^{-1} which, in spite of the smallness of m, may become very great as r diminishes; how small r may become we cannot say, its least possible value must depend on the extent to which the 'impenetrability' of matter is true of isolated molecules, for, since potential is only physically interpretable as the negative potential energy per unit mass of a particle (at least one molecule) at P, the least value of r is the least possible distance between the centres of two molecules. While not knowing this least value, we cannot but admit the possibility that a few terms of the type mr^{-1} might easily become so important as to make the potential quite different from the value of $\int \rho r^{-1} d\tau$ to which, as we shall see later, the parts of the hypothetical continuous distribution near P contribute only a negligible amount. But there is in any case, from the point of view of physics, no motive for pursuing the enquiry to such small values of r, for there are reasons for supposing that the Newtonian law of attraction does not hold good at such distances. In proving that the ordinary integral representations of potential and attractions are valid for distances of P from the attracting body which are indefinitely small from the point of view of molar physics, we have done all that would be required to justify their ordinary use in the theory of gravitation.

6. One point requires emphasis. By the attraction at a point P inside a body we mean the attraction (per unit mass) on a molecule or particle at P, situated in a cavity of dimensions which are only physically small. Hence if we describe a closed geometrical surface S, however small, in a body, we cannot calculate the force exerted on the part of the body inside S by the rest of the body by use of the attraction integrals. This can only be done when there is a gap, not smaller than physically small, between the attracting and the attracted matter, such a gap as might be made by cutting the body along the surface S and keeping the fissure open so that the opposite sides of it are nowhere in contact. In the absence of such a physical separation, account must be taken of unknown forces between molecules that are very near together. When the volume enclosed by S is not small beyond the physical limit of smallness, such molecules will all lie relatively near the surface S, and the forces between them will appear as surface forces between the geometrically separated portions of the body. In fluids the extra force is the fluid pressure, in solids it is less simple.

In the case of an absolutely continuous body there is nothing corresponding to the limit of physical smallness, and if the Newtonian law were supposed to hold for all distances however small there would be no surface forces of the kind described. The assumption of surface forces in the ideal case of continuity is really a tacit assumption that the Newtonian law breaks down ultimately as r diminishes.

7. We might, of course, devise other definite integrals than those above considered, in the hope of representing the same physical quantities with possibly greater accuracy. For example ρ might be obtained by averaging through a region smaller than physically small, so small that the number of molecules in it might be sometimes two or one or even zero; in this case fractions of molecules would become important, and the question would arise how a molecule ought to be regarded when it is neither altogether inside nor altogether outside the region. Again we might reduce the region of averaging to the limit of mathematical smallness and so get a ρ which is absolutely zero in intermolecular space, and presumably finite and continuous in the spaces occupied by the various molecules. Against such integrals it is to be urged firstly that one important factor of the function to be integrated, namely ρ, is inaccessible to experimental measurement, secondly that even if ρ were known the integrals would probably be more difficult to evaluate than the sum Σmr^{-1}, and thirdly that the greater accuracy which they seem to possess would be entirely vitiated by the probable failure of the Newtonian law for short distances. Moreover a method which involves integrating through the volumes of individual molecules, if it has any physical significance at all, implies the view that a molecule is of the nature of a small continuous mass whose smallest parts have the same kind of properties as the whole; this view is directly contrary to modern views of the constitution of matter, and the mathematical method corresponding to it, so far from being the best possible representation of the facts, must share all the defects of the method of summation for the various molecules.

II. Potentials and Attractions of accurately continuous bodies.

8. The potential and the attraction components of a finite body of accurately continuous substance, at an external point P, are represented by volume integrals which, for ordinary laws of density, give rise to no mathematical difficulties. The subjects of integration are finite at all points of the region of integration, and the integrals themselves are finite and differentiable with respect to the coordinates of P by the method known as 'differentiation under the sign of integration.' Thus the potential integral, defined as $\int \rho r^{-1} d\tau$, justifies its right to the name 'potential' by possessing the property that its differential coefficients with respect to the coordinates (ξ, η, ζ) of P are the attraction integrals of the type $\int \rho (x-\xi) r^{-3} d\tau$.

But it is quite another thing when we come to consider the potential and attractions at a point inside the gravitating body. For now, for example, if we define the potential as $\int \rho r^{-1} d\tau$ taken throughout the whole body, the subject of integration ρr^{-1} becomes infinite at the point P, a point in the volume of integration, and it becomes a question whether the integral symbol represents a finite quantity at all, and, if so, whether it is differentiable and what are its differential coefficients.

These troublesome questions might be avoided by introducing, as in the investigation for a body of molecular structure, a cavity within which P must be situated. And, indeed, this still seems to be demanded by the physical interpretation, since potential and attraction are physically defined as work function and force per unit mass for a hypothetical small mass or particle at the point P; such particle cannot be supposed to occupy space already occupied by other matter, and hence must be situated in a cavity made for it. But whereas, in the case of molecular structure, there was suggested from physical considerations a limit to the order of smallness of the cavity contemplated, corresponding in fact to the order of smallness of the necessarily present inaccuracy in the mathematical representation adopted, no such limit suggests itself in the case of continuous bodies. The retention of a cavity, of any definite though arbitrarily chosen order of smallness, is not demanded when there is no limit to the possible smallness of a portion of matter, and would moreover involve a want of precision or at least a restriction on the meaning of the mathematical symbols employed which would considerably discount their utility. Whereas it is only for the sake of mathematical precision that the hypothetical continuous bodies are generally made the subject of study in preference to the actual molecular bodies of which they are approximate representations.

9. We obtain the definiteness we desire, and, as will be seen, conform at the same time to the conventions and definitions of Integral Calculus, by framing new definitions of the potential and the attraction components at a point P (ξ, η, ζ), inside a continuous body. We first suppose the point P to be in a cavity, we then make the cavity smaller and smaller, and define the *limits* (if such exist) to which the potential and the attraction components at P tend with the vanishing of the cavity as the potential and the attraction components respectively at P when no cavity exists. It must be recognised that this passage to the limit entirely destroys the physical meaning which the quantities considered possess at any stage short of the limit, but on the other hand it gives us extremely convenient standard approximations to these quantities in cases of physical interest; the very definition of the term limit implies that the approximation can be made as close as we please by taking the cavity sufficiently small.

It is also to be noticed that a relation such as $X=\frac{\partial V}{\partial \xi}$ (where X is a force component and V the potential), which holds inside a cavity of finite size however small, might not persist after passage to the limit. That is to say, though of necessity

$$\operatorname{Lim} X=\operatorname{Lim} \frac{\partial V}{\partial \xi},$$

it is not equally inevitable that

$$\operatorname{Lim} X=\frac{\partial}{\partial \xi} \operatorname{Lim} V.$$

In fact if, as is customary, we drop the phrase 'limit' from our notation, though keeping the idea in mind, we have to face the fact that the formula $X=\frac{\partial V}{\partial \xi}$, valid for free space, requires examination before we can be sure that it is true at a point in the substance of the body. And if it be objected that the formula is known to be true in all cases of physical interest, and that no such interest attaches to its validity or otherwise in the case which has avowedly no physical significance, an answer is that if we decide to use a certain kind of mathematical functions as approximate representations of physical quantities, we must become acquainted with the meanings and properties of these functions before we can make intelligent use of them.

Hence it is natural for the student of the theory of attractions to turn his attention to that part of pure mathematics which has to do with the definition and properties of volume and surface integrals.

III. Volume integrals.

10. Let f be a function of position, and let a finite volume T be divided into a great number of elements $\Delta\tau$, of small linear dimensions; let f_1 be a quantity associated with an element of volume $\Delta\tau$, chosen according to some law, so that it is either the value of f at some point of the element, or at any rate not greater than the greatest or less than the least value of f for points in the element. If f is finite at all points in the volume T, the sum $\Sigma f_1 \Delta\tau$ extended to all elements of T is finite, and will remain so no matter how small and correspondingly numerous are the elements $\Delta\tau$. If this sum tends to a limit as the number of elements tends to infinity, and the linear dimensions of each tend to zero, and if this limit is independent of the law specifying f_1 and of the manner of subdivision into elements, the limit is called the volume integral of f through the volume T, and is denoted by $\int f d\tau$. This definition is only valid on the supposition that f is finite at all points in T.

Whether the limit here spoken of does or does not exist depends on the nature of the function f; we shall assume that it does exist for all the forms of f which we meet with in potential theory.

11. Next consider the case in which f is a function which becomes infinite at a point P within the volume T; clearly we need a new definition, and that which has been generally adopted is as follows. Surround the point P by a small closed surface t, and take the volume integral through the whole of the volume T except the part included by t; we thus exclude P from the range of integration, and so get a finite integral. Now let the surface t become smaller and smaller, whilst always surrounding P; if the volume integral tends to a finite limit as the space enclosed by t tends to vanishing, and if the limit is independent of the shape of t, then this limit is defined to be the integral of f throughout the whole volume T. The definition may be expressed symbolically thus:

$$\int^T f d\tau \equiv \operatorname*{Lim}_{t \to 0} \int_t^T f d\tau,$$

where the symbol $\to$ is used to denote such phrases as 'tending towards' or 'tends towards,' so that $t \to 0$ reads 'as t tends towards zero.' Here and elsewhere the subscript to the integral specifies the inner boundary of the region of integration.

If we call the space inside the vanishing surface t a 'cavity' in the volume of integration, we see at once the parallelism between the definition of this kind of volume integral and that of the so-called potential and attractions at a point in the substance of a continuous body.

The volume integral (if it exists) through a region within which f becomes infinite at some point is seen, by the above definition, to be a mathematical conception of a different character from the integral for a region in which f is everywhere finite. In a sense one might say that the latter is a true volume integral while the former is the limit of a true volume integral. The latter bears to the former the kind of relation that a single limit bears to a double limit, or that a finite series bears to the so-called sum of an infinite series.

12. Analogously with the terminology of series, we speak of the volume integral as convergent if it tends to a finite limit with the vanishing of the cavity, divergent if it tends to become infinitely great, and semi-convergent if, as sometimes happens, there is a finite limit whose value is not independent of the shape or mode of vanishing of the cavity. Divergent integrals are, for ordinary purposes, as meaningless as divergent infinite series, and so we must satisfy ourselves that the integrals in use in gravitation problems are convergent either absolutely or in the conditional manner corresponding to semi-convergence.

To decide whether, for a given form of f, the integral is convergent or not, we have the following rule, depending on the order of the infinity of f at P in terms of r, the distance from P to the point at which f is estimated. If f becomes infinite at P of an order lower than r^{-3} the volume integral is convergent, if of an order higher than r^{-3} the integral is divergent; if of the order r^{-3} exactly the integral may be divergent, semi-convergent, or convergent, according to the way in which f in the neighbourhood of P depends on the angular position of r. This rule is not stated with sufficient accuracy to rank as a theorem, and one can easily think of exceptions to it; for example the case of $f \equiv r^{-4} \cos\theta$ (in the notation of spherical polar coordinates), which is obviously convergent for any cavity symmetrical about the plane $\theta = \frac{1}{2}\pi$ though otherwise likely to be divergent, shews that something like semi-convergence may be associated with infinities of order greater than 3*; but the rule is a convenient approximation to the facts.

13. With a view to justifying the rule here given, it will be convenient to re-state, with slight modification of form, the definition of convergence. The integral of f through the volume T, which includes a point of infinity P, is convergent if, corresponding to any arbitrarily chosen small quantity σ, there can always be found a closed surface θ surrounding P such that all closed surfaces t surrounding P and lying wholly inside θ have the property that

$$\left| \int_t^\theta f d\tau \right| < \sigma.$$

That this is essentially the same as the definition of § 11 appears at once when we think of the ordinary definition of a limit; for if the integral through T has a limit A, we can choose θ so that

$$\left| \int_\theta^T f d\tau - A \right| < \tfrac{1}{2}\sigma, \quad \left| \int_t^T f d\tau - A \right| < \tfrac{1}{2}\sigma,$$

* See Article 70.

and therefore

$$\left|\int_t^\theta f\,d\tau\right| = \left|\int_t^T f\,d\tau - \int_\theta^T f\,d\tau\right| < \sigma\,;$$

it is, of course, to be understood that P must not lie on the surface θ. Thus the property constituting the definition of convergence of the present Article is a consequence of the property laid down as a definition in § 11.

Conversely, possession by an integral of the property specified in the present Article involves as a necessary consequence the existence of a limit A, though giving no indication of its actual value. For by taking θ sufficiently small we can keep the fluctuation of the value of the integral for different cavities within θ as small as we please, that is small without limit; and infinitely restricted fluctuation is the same as infinite approximation to some definite (and therefore finite) value.

Part of the rule of § 12 may be formulated in the following theorem. *If within a sphere of finite radius* (a), *having P as centre, f is everywhere less in absolute value than $Mr^{-\mu}$, where M is a definite constant and $\mu < 3$, the integral is convergent.* To prove this let us take for the surface θ the sphere $r = \eta$, where $\eta < a$, and let us denote by ϵ the distance from P to the nearest point of the surface t of the cavity; the cavity is of course entirely inside θ, but is otherwise unrestricted as to shape. Since the modulus of a sum is not greater than the sum of the moduli, and since an integral is the limit of a sum,

$$\begin{aligned}\left|\int_t^\theta f\,d\tau\right| &\leqslant \int_t^\theta |f|\,d\tau,\\ &\leqslant \int_\epsilon^\theta |f|\,d\tau,\\ &< M\int_\epsilon^\theta r^{-\mu}\,d\tau,\end{aligned}$$

where the subscript ϵ means that the inner boundary of the integrals is the sphere $r = \epsilon$; the second inequality holds because $|f|$ is positive and ϵ is completely inside θ.

In dealing with a function of r only, we may combine all elements $d\tau$ that lie between spheres of radii r and $r + dr$ in the single expression $4\pi r^2 dr$, so that

$$\begin{aligned}\left|\int_t^\theta f\,d\tau\right| &< 4\pi M\int_\epsilon^\eta r^{2-\mu}\,dr,\\ &< \frac{4\pi M}{3-\mu}(\eta^{3-\mu} - \epsilon^{3-\mu}),\\ &< \frac{4\pi M}{3-\mu}\eta^{3-\mu},\end{aligned}$$

it being noted that $\epsilon < \eta$ and that $3-\mu$ is positive, so that the last expression obtained is positive. Now if σ be any arbitrarily chosen small quantity, we have only to take η less than $\{(3-\mu)\sigma/4\pi M\}^{\frac{1}{3-\mu}}$ in order to get a surface θ such that

$$\left|\int_t^\theta f d\tau\right| < \sigma,$$

whatever shape t may have provided only it lies inside θ. Thus the convergence of the integral of f is established.

It will be noticed that the convergence of the integral of f in accordance with this theorem involves also, as the proof indicates, the convergence of the integral of $|f|$.

It need hardly be pointed out that the position and shape of the outer boundary T of the region of integration do not, in general, affect the question of convergence; whatever the outer boundary may be, provided it does not include other points of infinity, it is only the part of the volume just round P that is in danger of making the integral very great, and so only that part need be studied with a view to detecting divergence.

It is clear that the theorem holds equally well for cases in which the point P where the infinity occurs is not inside but just on the boundary of the region of integration.

14. The corresponding theorem for divergence is as follows. *If within a sphere of finite radius* (a), *having P as centre, f is everywhere algebraically greater than $mr^{-\mu}$, where m is a constant greater than zero, and $\mu \geqslant 3$, the integral is divergent.* To prove this, we take as outer boundary the sphere $r = a$, and as inner boundary a surface t, and we denote by ϵ the distance from P to the furthest point of the surface t, so that the sphere $r = \epsilon$ completely surrounds the cavity. Then

$$\int_t^a f d\tau > m \int_t^a r^{-\mu} d\tau > m \int_\epsilon^a r^{-\mu} d\tau;$$

as before, we collect all the elements $d\tau$ between r and $r + dr$ into the expression $4\pi r^2 dr$, and so get

$$\int_t^a f d\tau > 4\pi m \int_\epsilon^a r^{2-\mu} dr,$$

$$> \frac{4\pi m}{\mu - 3}\{\epsilon^{-(\mu-3)} - a^{-(\mu-3)}\} \text{ or } 4\pi m \{\log a - \log \epsilon\},$$

according as μ is greater than or equal to 3. In either case the expression obtained tends to infinity for $\epsilon \to 0$; and so the integral, being greater than a quantity which tends to become endlessly great, is divergent.

15. The case of semi-convergence need only be illustrated by an example. Suppose that f is $r^{-3}\cos\theta$, and consider the values of the integral in the regions having a common outer boundary $r=a$, and having for inner boundaries in the first instance the sphere $r=\epsilon$, and in the second instance the sphere $r^2 - r\epsilon\cos\theta = 2\epsilon^2$, these being two small spheres of which the latter touches and completely surrounds the former.

The difference between the integrals over these two regions is the integral through the space between the two small spheres, which is certainly not zero so long as ϵ is different from zero, since, if we consider the volumes of the region cut off by a cone of small solid angle having the origin as vertex, the positive contribution to the integral from the frustum where $\cos\theta$ is positive is greater in absolute value than the negative contribution from the frustum where $\cos\theta$ is negative. Further, the magnitude of the integral is independent of ϵ, since if we multiply ϵ by k we can get the new region of integration by multiplying all radii vectores from P by k, and thus each element of volume $d\tau$ is multiplied by k^3; the subject of integration $r^{-3}\cos\theta$ is correspondingly multiplied by k^{-3}, and so the integral is unaltered. Thus the integral over the space between the small spheres, being finite when ϵ is not zero, has the same finite value as ϵ tends to vanishing; in other words there is a finite difference between the values of the original integral corresponding to the different cavities. It is clear, from the symmetry of f about the plane $\theta = \frac{1}{2}\pi$, that the integral is zero for the cavity whose centre is P, and therefore not zero for the cavity whose centre is not at P; but in neither case is it infinite. Thus the semi-convergence of the integral is demonstrated.

This example suggests the remark that two cavities are to be regarded as of different shapes even if they are similar, if they are not similarly situated with respect to P. The cavities considered in the example are both spheres, but since one has P at its centre while in the other P trisects a diameter, the cavities are regarded as of different shapes for purposes of the present discussion.

IV. Theorems connecting volume and surface integrals.

16. There is a well-known theorem connecting volume integrals with surface integrals taken over the boundary of the region of volume-integration. If the region be finite, if l, m, n denote the direction cosines of the normal drawn outwards from the region at a point of the boundary B, and if ξ, η, ζ be functions having finite space differential coefficients at all points in the region,

$$\int (l\xi + m\eta + n\zeta)\, dS = \int \left(\frac{\partial \xi}{\partial x} + \frac{\partial \eta}{\partial y} + \frac{\partial \zeta}{\partial z}\right) d\tau \;\ldots\ldots\ldots\ldots(1),$$

where dS represents an element of area of the boundary, the surface integral is taken over the complete boundary, and the volume integral through the complete volume. A proof is given in Williamson's *Integral Calculus*, Chapter XI.

It is worth while to enquire whether this theorem can be extended to the case in which there is a point P in the volume where the subject of volume-integration becomes infinite. The course which suggests itself is to surround the point P by a small closed surface σ, and to apply the original theorem to the region bounded internally by σ and externally by the surface B. The complete boundary consists of both B and σ, and so we get

$$\int_\sigma^B \left(\frac{\partial \xi}{\partial x} + \frac{\partial \eta}{\partial y} + \frac{\partial \zeta}{\partial z}\right) d\tau = \int_B (l\xi + m\eta + n\zeta)\, dS + \int_\sigma (l\xi + m\eta + n\zeta)\, dS,$$

where the suffixes to the surface integrals indicate the surfaces over which they are taken. Now if the subject of integration of the volume integral is such as to make it convergent with respect to the infinity at P, the left-hand side of the equality tends to a definite limit as the dimensions of σ decrease towards zero. Consequently we get as the limiting form of the equality,

$$\int^B \left(\frac{\partial \xi}{\partial x} + \frac{\partial \eta}{\partial y} + \frac{\partial \zeta}{\partial z}\right) d\tau$$

$$= \int_B (l\xi + m\eta + n\zeta)\, dS + \lim_{\sigma \to 0} \int_\sigma (l\xi + m\eta + n\zeta)\, dS \quad \ldots(2).$$

When the volume integral is convergent, the left side of (2) is a perfectly definite finite quantity, and hence the limit indicated on the right-hand side must exist and be independent of the shape of σ; it will exist, but have a value dependent on the shape of σ, if the volume integral is semi-convergent. In the former case it is frequently convenient to determine the value of the limit by taking some specially simple shape for σ, such as a sphere with centre at P. Generally, if the subject of volume-integration is, in the neighbourhood of P, of order $r^{-\mu}$ $(\mu < 3)$, where r denotes distance from P, the subject of the surface integral under the limit sign is of order $r^{-\mu+1}$, where r equals the radius ϵ of the sphere σ; also dS is $\epsilon^2 d\omega$, where $d\omega$ is an element of solid angle, and so the surface integral is of order $\epsilon^{3-\mu}$ and tends to the limit zero for $\epsilon \to 0$. If $\mu = 3$, the case of possible semi-convergence, the surface integral is of order ϵ^0, so far as its dependence on ϵ is concerned, and therefore may have for limit a value different from zero.

17. A generalisation, and at the same time a particular case, of the fundamental 'surface and volume integral theorem' is got by putting $\phi\xi$, $\phi\eta$, $\phi\zeta$, instead of ξ, η, ζ, where ϕ is another function of position which has finite space differential coefficients at all points in the region. The volume integral then becomes

$$\int \left\{ \frac{\partial}{\partial x}(\phi\xi) + \frac{\partial}{\partial y}(\phi\eta) + \frac{\partial}{\partial z}(\phi\zeta) \right\} d\tau,$$

so that the theorem takes the form

$$\int \phi \,.\, (l\xi + m\eta + n\zeta)\, dS - \int \phi \,.\left(\frac{\partial \xi}{\partial x} + \frac{\partial \eta}{\partial y} + \frac{\partial \zeta}{\partial z}\right) d\tau$$
$$= \int \left(\xi \frac{\partial \phi}{\partial x} + \eta \frac{\partial \phi}{\partial y} + \zeta \frac{\partial \phi}{\partial z} \right) d\tau \;\ldots\ldots\ldots(3).$$

Now ϕ is continuous and therefore finite throughout the whole region of integration, but let us suppose that some or all of the functions ξ, η, ζ, $\frac{\partial \xi}{\partial x}$, $\frac{\partial \eta}{\partial y}$, $\frac{\partial \zeta}{\partial z}$ become infinite at a point P in the region. If the infinities are such that both the volume integrals are convergent, it is clear that by introducing the cavity σ and making it tend to zero dimensions we get the relation

$$\int_B \phi \,.\, (l\xi + m\eta + n\zeta)\, dS + \operatorname*{Lim}_{\sigma \to 0} \int_\sigma \phi \,.\, (l\xi + m\eta + n\zeta)\, dS$$
$$- \int^B \phi \,.\left(\frac{\partial \xi}{\partial x} + \frac{\partial \eta}{\partial y} + \frac{\partial \zeta}{\partial z}\right) d\tau$$
$$= \int^B \left(\xi \frac{\partial \phi}{\partial x} + \eta \frac{\partial \phi}{\partial y} + \zeta \frac{\partial \phi}{\partial z} \right) d\tau \;\ldots\ldots\ldots\ldots\ldots\ldots(4),$$

and that when the convergence is due to ξ, η, ζ being infinite of lower order than r^{-2} the limit of the surface integral is zero. When the volume integrals are semi-convergent, or when ξ, η, ζ are infinite of the same order as r^{-2}, the limit of the surface integral may be different from zero, possibly depending on the shape of the cavity; it would usually be convenient to take it in the ultimately equivalent form

$$\phi_P \,.\, \operatorname*{Lim}_{\sigma \to 0} \int_\sigma (l\xi + m\eta + n\zeta)\, dS \;\ldots\ldots\ldots\ldots\ldots\ldots(5),$$

where ϕ_P signifies the value of ϕ at the point P.

18. Green's Theorem is got from the 'surface and volume integral theorem' by putting

$$\xi = U\frac{\partial V}{\partial x}, \qquad \eta = U\frac{\partial V}{\partial y}, \qquad \zeta = U\frac{\partial V}{\partial z},$$

where U, V are functions of position; if we use the notation $\frac{\partial}{\partial\nu}$ for differentiation along the outward normal, and Δ for Laplace's operator $\frac{\partial^2}{\partial x^2}+\frac{\partial^2}{\partial y^2}+\frac{\partial^2}{\partial z^2}$, we get

$$\int U\frac{\partial V}{\partial\nu}dS = \int\Sigma\left\{\frac{\partial}{\partial x}\left(U\frac{\partial V}{\partial x}\right)\right\}d\tau,$$

whence

$$\int U\frac{\partial V}{\partial\nu}dS - \int U\Delta V d\tau = \int\Sigma\left(\frac{\partial U}{\partial x}\frac{\partial V}{\partial x}\right)d\tau$$

$$= \int V\frac{\partial U}{\partial\nu}dS - \int V\Delta U d\tau \quad \text{............(6)}$$

by symmetry. These two equalities constitute Green's Theorem. The statement of the theorem requires modification when the region of integration is multiply connected and either U or V is many-valued; this modification is discussed in Maxwell's *Electricity* and in Lamb's *Hydrodynamics*, and need not be entered upon here.

If one of the functions, say U, together with all of its differential coefficients which occur in the formula (6), is finite throughout the region, and if V becomes infinite at a point P in the region, we isolate P in a cavity σ and make the dimensions of σ tend to zero. If the volume integrals converge, we get a pair of equalities precisely similar to the above, save that to the first member we must add $\lim_{\sigma\to 0}\int_\sigma U\frac{\partial V}{\partial\nu}dS$, and to the third member $\lim_{\sigma\to 0}\int_\sigma V\frac{\partial U}{\partial\nu}dS$.

Generally the convergence of all the volume integrals would involve that V should become infinite at P of an order lower than r^{-1}, since when V is a function of position relative to P, as is frequently the case, a space differentiation adds one to the order of the infinity so that ΔV is of an order higher by r^{-2} than V. In this case both the integrals over the surface σ are of the order of a positive power of the small length r, and their limits are zero.

The case of $V = r^{-1}$ is one of special interest in physical applications; it is also interesting mathematically because it is just the case in which semi-convergence is to be looked for. There is no semi-convergence however, for $\int U\Delta(r^{-1})\,d\tau$ is absolutely zero, since

$\Delta(r^{-1}) = 0$ at all points between σ and the outer boundary, and the other volume integrals are convergent because the subjects of integration are infinite of lower order than r^{-3}; and if all the terms but one of an equality are definite or convergent, that one cannot be semi-convergent. Clearly, since dS is comparable with $r^2 d\omega$, where $d\omega$ is an element of solid angle, $\int_\sigma r^{-1} \frac{\partial U}{\partial \nu} dS$ has a zero limit; but $\int_\sigma U \frac{\partial}{\partial \nu}(r^{-1})\, dS$ (when σ is a sphere of radius ϵ, which, in the absence of semi-convergence, may be assumed without loss of generality) has the same limit as

$$U_P \int_\sigma \frac{\partial}{\partial \nu}(r^{-1})\, r^2 d\omega \quad \text{or} \quad -U_P \int_\sigma \frac{\partial}{\partial r}(r^{-1})\, r^2 d\omega,$$

that is $U_P \int d\omega$ or $4\pi U_P$.

Thus Green's Theorem in this case gives the equalities

$$\int U \frac{\partial}{\partial \nu}(r^{-1})\, dS + 4\pi U_P = \int \Sigma \left\{\frac{\partial U}{\partial x} \frac{\partial}{\partial x}(r^{-1})\right\} d\tau$$

$$= \int r^{-1} \frac{\partial U}{\partial \nu} dS - \int r^{-1} \Delta U d\tau \quad \text{............(7).}$$

Almost identical reasoning applies to the case in which V satisfies $\Delta V = 0$ at all points of the region other than P, and becomes infinite at P in such a way that $\operatorname{Lim}_{r \to 0} (rV) = M$, where M is a definite constant. The theorem becomes

$$\int U \frac{\partial V}{\partial \nu} dS + 4\pi M U_P = \int \Sigma \left(\frac{\partial U}{\partial x} \frac{\partial V}{\partial x}\right) d\tau$$

$$= \int V \frac{\partial U}{\partial \nu} dS - \int V \Delta U d\tau.$$

Another case of Green's Theorem much used in Physics is that in which $U = V$. The two equalities reduce to the single one

$$\int V \frac{\partial V}{\partial \nu} dS - \int V \Delta V d\tau = \int \Sigma \left(\frac{\partial V}{\partial x}\right)^2 d\tau \text{............(8);}$$

no particular interest attaches to an examination of possible modifications in this formula when V has an infinity at a point in the region of integration.

V. The differentiation of volume integrals.

19. We shall next discuss the possibility of differentiating a volume integral with respect to a parameter which occurs in the subject of integration but does not affect the boundary of the region of integration. The only parameters that need be considered here are the coordinates of a point P at which the subject of integration f has an infinity. We shall call these coordinates ξ, η, ζ, keeping x, y, z to denote the coordinates of the element $d\tau$ of integration. The question to be settled is whether the integral has a differential coefficient with respect to ξ, and if so whether that differential coefficient is equal to the integral of $\partial f/\partial \xi$. Integration means passage to a limit, and so also does differentiation; we have to find out whether alteration of the order of the two passages to limit alters the value of the final result. When P is in the region of integration there is in each case the additional passage to limit corresponding to the closing of the cavity, and the question to be settled is whether by first integrating, second closing the cavities, third making $\Delta \xi \to 0$, (where $\Delta \xi$ is an increment of ξ), we get the same result as by first making $\Delta \xi \to 0$, second integrating, third closing the cavity.

20. First we shall consider the case in which P is outside the region of integration, and shew that if f has at all points of the region and for all contemplated values of ξ a differential coefficient with respect to ξ, which differential coefficient is a uniformly continuous* function of ξ throughout the region, the differential coefficient of the integral is the same as the integral of the differential coefficient.

The incremental ratio of the integral is

$$\frac{1}{\Delta \xi}\int \{f(\xi + \Delta \xi) - f(\xi)\}\, d\tau,$$

which, by the theorem of mean value,

$$= \int f'(\xi + \theta \Delta \xi)\, d\tau$$
$$= \int f'(\xi)\, d\tau + \int \epsilon d\tau,$$

where $\epsilon \equiv f'(\xi + \theta \Delta \xi) - f'(\xi)$, and $1 > \theta > 0$.

Now the uniform continuity of $f'(\xi)$ implies that for an arbitrarily chosen small quantity σ we can always find a quantity ω such that for all values of $\Delta \xi$ less than ω,

$$|f'(\xi + \Delta \xi) - f'(\xi)|,$$

and consequently also $|\epsilon|$, is less than σ, for all points in the region T of integration. Thus by choosing $\Delta \xi$ less than ω we can ensure that $|\int \epsilon d\tau|$ shall be less than σT, which, in virtue of the finiteness of T and the arbitrariness of σ, is arbitrarily small. In fact the difference between the

* It can be proved that if a function is continuous at all points in a region it is *uniformly* continuous throughout the region.

integral of $f'(\xi)$ and the incremental ratio of the integral of $f(\xi)$ can be made arbitrarily small. Hence the limit of the incremental ratio, i.e. $\frac{\partial}{\partial \xi}\int f(\xi)\, d\tau$, is equal to $\int f'(\xi)\, d\tau$, as we set out to prove.

21. Next we consider the case in which P is within the region of integration. The rough rule for this case is that if the original integral is convergent, and if the integral obtained by differentiating under the sign of integration is convergent, the latter is the differential coefficient of the former.

It will serve our purpose to prove this proposition for a particular case, namely that in which the subject of integration f is the product of two factors, each subject to special restrictions. One of these, which we denote by $\rho(x, y, z)$ or briefly by ρ, is supposed to be a function of absolute position, not involving ξ, η, ζ at all; it is assumed to be finite, not to vanish at P, and to have space differential coefficients which are uniformly continuous throughout the region of integration. The other factor is supposed to be a function of position relative to P, and to become infinite at P; we may denote it by $\phi(x-\xi, y-\eta, z-\zeta)$, or sometimes for brevity by $\phi(\xi, x)$ or $\phi(\xi)$. It has obviously the property that $\phi(\xi + \Delta\xi, x) = \phi(\xi, x - \Delta\xi)$, so that

$$\frac{\partial \phi}{\partial \xi} = -\frac{\partial \phi}{\partial x}.$$

The integral to be differentiated is $\int^T \rho \,.\, \phi \,.\, d\tau$, or, written in full, $\operatorname*{Lim}_{\epsilon \to 0} \int_\epsilon^T \rho \,.\, \phi \,.\, d\tau$, where ϵ is a cavity surrounding the point P. The incremental ratio is

$$\frac{1}{\Delta\xi}\left[\operatorname*{Lim}_{\epsilon' \to 0} \int_{\epsilon'}^T \rho(x, y, z) \,.\, \phi(\xi + \Delta\xi)\, d\tau - \operatorname*{Lim}_{\epsilon \to 0} \int_\epsilon^T \rho(x, y, z) \,.\, \phi(\xi)\, d\tau\right],$$

where ϵ' is a cavity surrounding the point $P'(\xi + \Delta\xi, \eta, \zeta)$, which may be taken to be in all respects similar to ϵ. To get the differential coefficient of the integral it is necessary, in the incremental ratio, first to make ϵ and ϵ' tend to vanishing, and afterwards to make $\Delta\xi \to 0$.

Before we pass to either limit, however, we are at liberty to simplify the form of the incremental ratio in any manner that seems desirable. In the integral $\int_{\epsilon'}^T \rho(x, y, z) \,.\, \phi(\xi + \Delta\xi)\, d\tau$ imagine the boundaries of the region of integration, and every volume element, shifted a distance $\Delta\xi$ in the negative direction of the axis of x. An element of volume originally at a point K' is merely shifted to a point K whose position relative to the point (ξ, η, ζ) is the same as that of K' relative to $(\xi + \Delta\xi, \eta, \zeta)$; so that if K is (x, y, z), the value of ϕ belonging to the shifted volume element, originally presenting itself as $\phi(x + \Delta\xi, \xi + \Delta\xi)$, is equally well represented by the form $\phi(x, \xi)$; but the value of ρ for the element now brought to K is that appropriate to K', i.e. $\rho(x + \Delta\xi, y, z)$. The inner boundary ϵ' of the integral is brought by the shifting into coincidence with ϵ, but the outer boundary is changed to a surface T' which is simply T displaced without change of form. In fact

$$\int_{\epsilon'}^{T} \rho(x) \,.\, \phi(\xi + \Delta\xi, x)\, d\tau = \int_{\epsilon}^{T'} \rho(x + \Delta\xi) \,.\, \phi(\xi, x)\, d\tau,$$

so that the incremental ratio of the integral with as yet unclosed cavity is equal to

$$\frac{1}{\Delta\xi}\left[\int_{\epsilon}^{T'} \rho(x + \Delta\xi) \,.\, \phi(\xi)\, d\tau - \int_{\epsilon}^{T} \rho(x) \,.\, \phi(\xi)\, d\tau\right]$$

or to

$$\frac{1}{\Delta\xi}\left[\int_{\epsilon}^{T} \{\rho(x + \Delta\xi, y, z) - \rho(x, y, z)\} \,.\, \phi(\xi)\, d\tau + \int_{T}^{T'} \rho(x + \Delta\xi, y, z) \,.\, \phi(\xi)\, d\tau\right]$$

where the latter integral is extended to the region between T and T', being taken positively where the boundary of T' lies outside T, negatively where the reverse is the case. By the theorem of mean value the above expression equals

$$\int_{\epsilon}^{T} \rho_x'(x + \theta\Delta\xi, y, z) \,.\, \phi(\xi)\, d\tau + \frac{1}{\Delta\xi}\int_{T}^{T'} \rho(x + \Delta\xi) \,.\, \phi(\xi)\, d\tau,$$

$$= \int_{\epsilon}^{T} \rho_x'(x, y, z) \,.\, \phi(\xi)\, d\tau + \int_{\epsilon}^{T} \omega\,\phi(\xi)\, d\tau + \frac{1}{\Delta\xi}\int_{T}^{T'} \rho(x + \Delta\xi) \,.\, \phi(\xi)\, d\tau,$$

where $\omega \equiv \rho_x'(x + \theta\Delta\xi, y, z) - \rho_x'(x, y, z)$, and $1 > \theta > 0$.

Now as P is not on the boundary of T, $\Delta\xi$ can always be taken small enough to prevent P from being in the region between T and T'; hence the integral for this region has no dependence on the cavity ϵ. The assumed uniform continuity (which includes finiteness) of ρ_x', and the assumed convergence of the original integral, are sufficient guarantees of the convergence of the integrals of $\rho_x'\phi$ and $\omega\phi$. Hence we may proceed to the limits for $\epsilon \to 0$ and $\epsilon' \to 0$, and so get the relation:

Incremental ratio of $\int^{T} \rho\phi\, d\tau$

$$= \int^{T} \rho_x'\phi\, d\tau + \int^{T} \omega\phi\, d\tau + \frac{1}{\Delta\xi}\int_{T}^{T'} \rho(x + \Delta\xi) \,.\, \phi(\xi)\, d\tau \quad \ldots(9),$$

and the cavities ϵ, ϵ' are now closed up and finished with.

As $\Delta\xi$ becomes smaller it is clear that the volume between T and T' approximates to a very thin shell over the surface of T whose normal thickness (outwards from T) is $-\Delta\xi . l$, where l is the x cosine of the outward normal. Thus the corresponding volume integral approximates to and has the same limit as the surface integral

$$-\Delta\xi\int_T l\rho\phi\, dS.$$

And in virtue of the uniform continuity of ρ_x', corresponding to any arbitrary small quantity σ, we can always find a quantity κ such that for all values of $\Delta\xi$ less than κ, and for all points of the region of integration, $|\omega| < \sigma$, and so

$$\left|\int^T \omega\phi\, d\tau\right| < \sigma\int^T |\phi|\, d\tau.$$

This last expression is σ multiplied by a finite quantity, for, since ρ is finite and does not vanish at P, the convergence of $\int^T |\phi|\, d\tau$ is implied in the assumed convergence of $\int^T \rho\phi\, d\tau$, at least if the latter be convergence of the kind discussed in § 13. Hence $\int^T \omega\phi\, d\tau$ can, by suitable choice of $\Delta\xi$, be made smaller than any arbitrary small quantity, and so tends to the limit zero for $\Delta\xi \to 0$. Proceeding now to the limit $\Delta\xi \to 0$ in relation (9), we get

$$\frac{\partial}{\partial\xi}\int^T \rho\phi\, d\tau = \int^T \rho_x'\phi\, d\tau - \int_T l\rho\phi\, dS \quad \text{..........(10)}.$$

The theorem of § 17, formula (4), enables us to transform the right-hand side of (10), giving

$$\frac{\partial}{\partial\xi}\int^T \rho\phi\, d\tau = -\int^T \rho\phi_x'\, d\tau$$

$$= \int^T \rho\frac{\partial\phi}{\partial\xi}\, d\tau$$

$$= \int^T \frac{\partial}{\partial\xi}(\rho\phi)\, d\tau \quad \text{................(11)},$$

provided the last integral is convergent.

If the function ρ is of such a simple character near P that the nature of the integral depends entirely on the form of ϕ, it is clear that the case of possible semi-convergence is covered by the above reasoning, certainly as far as formula (10); but in formula (11) the use of formula (4) may introduce an extra term on the right-hand side, namely the limit of the surface integral of $l\rho\phi$ over a new cavity round P. One can imagine that peculiarities in the form of ρ might invalidate some of the steps of the argument, but such peculiarities are not to be expected in physical applications.

21 a. The more general problem of the differentiation of a volume integral with respect to a parameter θ which affects the boundary of the region of integration as well as the subject of integration may be mentioned here. The formula for the differentiation is

$$\frac{d}{d\theta}\int f(x, y, z, \theta)\, d\tau = \int f_\theta'(x, y, z, \theta)\, d\tau - \int f(x, y, z, \theta)\frac{\partial F}{\partial \theta}\left[\Sigma\left(\frac{\partial F}{\partial x}\right)^2\right]^{-\frac{1}{2}} dS,$$

where the region of volume integration is bounded by the surface $F(x, y, z, \theta) = 0$, dS is an element of area on this surface, and f' is the first derived function of f. When it is assumed that the subjects of both integrations on the right-hand side are uniformly continuous in their respective regions of integration the proof presents no difficulty and may be left to the reader. Infinities of f would require special investigation and might introduce exceptions to the formula.

VI. Applications to Potential Theory.

22. The potential at a point $P(\xi, \eta, \zeta)$ of a finite mass of continuous matter, whose density at a point (x, y, z) is ρ, a function of x, y, z but not of ξ, η, ζ, is the volume integral $V \equiv \int \rho r^{-1} d\tau$, where $r = \sqrt{\Sigma (x-\xi)^2}$; the attraction component parallel to the axis of x is X where $X \equiv \int (x-\xi) r^{-3} d\tau$; both integrals are taken through the whole space occupied by the body.

When (ξ, η, ζ) is outside the body, and ρ is finite and subject to such restrictions as are required for the validity of the theorems proved in the preceding Articles, there is no infinity of the subjects of integration in the region, and so

$$X = \int \frac{\partial}{\partial \xi}(\rho r^{-1})\, d\tau = \frac{\partial}{\partial \xi}\int \rho r^{-1}\, d\tau = \frac{\partial V}{\partial \xi}.$$

Also
$$\frac{\partial^2 V}{\partial \xi^2} = \frac{\partial}{\partial \xi}\int \frac{\partial}{\partial \xi}(\rho r^{-1})\, d\tau = \int \frac{\partial^2}{\partial \xi^2}(\rho r^{-1})\, d\tau = \int \rho \frac{\partial^2}{\partial \xi^2}(r^{-1})\, d\tau,$$

so that
$$\frac{\partial^2 V}{\partial \xi^2} + \frac{\partial^2 V}{\partial \eta^2} + \frac{\partial^2 V}{\partial \zeta^2} = \int \rho\left(\frac{\partial^2}{\partial \xi^2} + \frac{\partial^2}{\partial \eta^2} + \frac{\partial^2}{\partial \zeta^2}\right)(r^{-1})\, d\tau = 0.$$

23. When the point P is inside the body, the potential and the attraction components have no longer the simple physical interpretation suggested by their names, but are defined as the limits of these physical quantities in a vanishing cavity. And this, as we saw in § 11, implies their equivalence to the integrals represented by the same symbols as in § 22, but referring to a region including the point of infinity P of the subjects of integration, and therefore only intelligible when the integrals are convergent.

The subject of the potential integral, being infinite at P of the order of r^{-1}, the integral is convergent; and the attraction integrals, having subjects of integration that are infinite of the order of r^{-2}, are also convergent. Hence we have, by § 21,

$$X = \frac{\partial V}{\partial \xi}.$$

But if we differentiate X with respect to ξ under the sign of integration, we get an integral whose subject of integration is of the order of r^{-3}, so that there is a possibility of semi-convergence or divergence. · Instead, therefore, of merely quoting a simple differentiation rule in this case, we must proceed with care.

It is clear that for the integral $X \equiv \int \rho\,(x-\xi)\,r^{-3}\,d\tau$ the argument of § 21 holds as far as the formula (10), which in this case becomes

$$\frac{\partial X}{\partial \xi} = \int^T \frac{\partial \rho}{\partial x}(x-\xi)\,r^{-3}d\tau - \int_T l\rho\,(x-\xi)\,r^{-3}dS \quad \text{......(12)},$$

the volume integrals involved being convergent and all cavities being closed up. This formula shews that $\dfrac{\partial X}{\partial \xi}$ has a definite value.

Now surround P by a small surface σ and use formula (4) of § 17, putting ρ for the quantity there called ϕ, and $(x-\xi)\,r^{-3}$ for the quantity there called ξ. Thus we get

$$\int_T l\rho\,(x-\xi)\,r^{-3}dS + \rho_P \operatorname*{Lim}_{\sigma\to 0}\int_\sigma l\,(x-\xi)\,r^{-3}dS$$
$$-\operatorname*{Lim}_{\sigma\to 0}\int_\sigma^T \rho\,\frac{\partial}{\partial x}\{(x-\xi)\,r^{-3}\}\,d\tau = \int^T \frac{\partial \rho}{\partial x}(x-\xi)\,r^{-3}\,d\tau,$$

whence

$$\frac{\partial X}{\partial \xi} = \rho_P \operatorname*{Lim}_{\sigma\to 0}\int_\sigma l\,(x-\xi)\,r^{-3}dS - \operatorname*{Lim}_{\sigma\to 0}\int_\sigma^T \rho\,\frac{\partial}{\partial x}\{(x-\xi)\,r^{-3}\}\,d\tau \quad \text{...(13)}.$$

The sum of the limits here indicated is perfectly definite and independent of the shape of σ, but either limit taken separately has a value which depends on the shape of σ; this can be seen readily by studying the surface integral first when σ is a sphere and second when σ is a very flat cylinder with plane ends parallel to the plane of x.

Since formulae corresponding to (13) hold for Y and Z, we get by addition

$$\frac{\partial X}{\partial \xi} + \frac{\partial Y}{\partial \eta} + \frac{\partial Z}{\partial \zeta} = \rho_P \operatorname*{Lim}_{\sigma \to 0} \int_\sigma \Sigma l\,(x-\xi)\, r^{-3} dS$$
$$- \operatorname*{Lim}_{\sigma \to 0} \int_\sigma^T \rho \Sigma \frac{\partial}{\partial x} \{(x-\xi)\, r^{-3}\}\, d\tau.$$

Now here the subject of volume-integration is identically zero at all points outside the cavity, and so the integral is zero whatever the shape of the cavity, and its limit is zero. Hence the value of the surface integral is independent of the shape of the cavity, and may be calculated on the assumption that σ is the sphere $r = \epsilon$; in this case $l = -(x-\xi)\,\epsilon^{-1}$, so that $\Sigma l\,(x-\xi) = -\epsilon$, and $dS = \epsilon^2 d\omega$, where $d\omega$ is an element of solid angle at P; thus the integral becomes $-\int d\omega$, which equals -4π. Thus our equality becomes

$$\frac{\partial X}{\partial \xi} + \frac{\partial Y}{\partial \eta} + \frac{\partial Z}{\partial \zeta} = -4\pi\rho_P \quad \text{..............} (13\,a),$$

or

$$\frac{\partial^2 V}{\partial \xi^2} + \frac{\partial^2 V}{\partial \eta^2} + \frac{\partial^2 V}{\partial \zeta^2} = -4\pi\rho_P,$$

which is Poisson's equation.

24. The theorems proved above for the differential coefficients of V are perfectly intelligible for a body of the hypothetical continuous structure which we have postulated. But when applied to a body of molecular structure such a symbol as $\frac{\partial V}{\partial \xi}$ requires qualification. It has been seen that for a continuous body $\Delta\xi$ was only made to tend to zero after we had first closed the cavities ϵ and ϵ' corresponding to V and $V + \Delta V$; in fact $\Delta\xi$, though small, was always large compared with the dimensions of the cavities, but this fact did not interfere with our making $\Delta\xi$ as near to zero as we pleased.

But for a body of molecular structure the cavities must always be large enough to be capable of containing a great number of molecules, and so we can never close them entirely; hence $\Delta\xi$, so far from ever vanishing, must be actually large compared with the smallest length which can be regarded as only physically small; nevertheless $\Delta\xi$ may be, to our senses, extremely small. Hence instead of the true differential coefficient $\frac{\partial V}{\partial \xi}$ we have the incremental ratio $\frac{\Delta V}{\Delta \xi}$, where $\Delta\xi$ though very small is still definitely prevented from attaining the higher orders of smallness which lie on the way to the limit zero. Thus it is clear that the symbol $\frac{\partial V}{\partial \xi}$, just as V itself, is inexact and stands for something not precisely defined; but the inaccuracy or deviation from a precise value is no greater than the inaccuracy which regards matter as continuous, and is in fact an inaccuracy so small as to be inappreciable to our senses. Accordingly the relations

$$X = \frac{\partial V}{\partial \xi} \quad \text{and} \quad \Sigma \frac{\partial X}{\partial \xi} = -4\pi\rho$$

have a sufficiently precise meaning when applied to bodies of molecular structure.

VII. Applications to Theory of Magnetism.

25. If a body is magnetised so that the components of the intensity of magnetisation at a point (x, y, z) are A, B, C, the magnetic potential at an external point P, (ξ, η, ζ), is given by

$$V = \int \left(A \frac{\partial}{\partial x} + B \frac{\partial}{\partial y} + C \frac{\partial}{\partial z} \right) (r^{-1})\, d\tau \quad \text{............(14)},$$

and the x component of magnetic force is α where

$$\alpha = -\int \frac{\partial}{\partial \xi} \left(A \frac{\partial}{\partial x} + B \frac{\partial}{\partial y} + C \frac{\partial}{\partial z} \right) (r^{-1})\, d\tau \quad \text{........(15)},$$

the body being regarded as of continuous structure; so long as P is outside the body it is clear that

$$\alpha = -\frac{\partial V}{\partial \xi} \quad \text{..........................(16)}.$$

Outside the body the induction (a, b, c) is the same as the force (α, β, γ), and therefore remembering that A, B, C are functions of x, y, z but not of ξ, η, ζ, while r depends only on $x-\xi$, $y-\eta$, $z-\zeta$, we see that

$$\begin{aligned}
a &= -\int \left(A \frac{\partial}{\partial x} + B \frac{\partial}{\partial y} + C \frac{\partial}{\partial z} \right) \frac{\partial}{\partial \xi} (r^{-1})\, d\tau \\
&= \int \left(A \frac{\partial^2}{\partial x^2} + B \frac{\partial^2}{\partial x \partial y} + C \frac{\partial^2}{\partial x \partial z} \right) (r^{-1})\, d\tau \\
&= \int \left(-A \frac{\partial^2}{\partial y^2} - A \frac{\partial^2}{\partial z^2} + B \frac{\partial^2}{\partial x \partial y} + C \frac{\partial^2}{\partial x \partial z} \right) (r^{-1})\, d\tau \\
&= \int \left(C \frac{\partial}{\partial x} - A \frac{\partial}{\partial z} \right) \frac{\partial}{\partial z} (r^{-1})\, d\tau - \int \left(A \frac{\partial}{\partial y} - B \frac{\partial}{\partial x} \right) \frac{\partial}{\partial y} (r^{-1})\, d\tau \\
&= -\int \left(C \frac{\partial}{\partial x} - A \frac{\partial}{\partial z} \right) \frac{\partial}{\partial \zeta} (r^{-1})\, d\tau + \int \left(A \frac{\partial}{\partial y} - B \frac{\partial}{\partial x} \right) \frac{\partial}{\partial \eta} (r^{-1})\, d\tau \\
&= \frac{\partial H}{\partial \eta} - \frac{\partial G}{\partial \zeta} \quad \text{..(17)},
\end{aligned}$$

where

$$F = \int\left(B\frac{\partial}{\partial z} - C\frac{\partial}{\partial y}\right)(r^{-1})\,d\tau, \qquad G = \int\left(C\frac{\partial}{\partial x} - A\frac{\partial}{\partial z}\right)(r^{-1})\,d\tau,$$

$$H = \int\left(A\frac{\partial}{\partial y} - B\frac{\partial}{\partial x}\right)(r^{-1})\,d\tau \quad\ldots\ldots\ldots\ldots\ldots\ldots(18).$$

F, G, H are the components of the vector potential at P; the relation between induction and vector potential is frequently written

$$(a, b, c) = \text{curl}\,(F, G, H) \quad\ldots\ldots\ldots\ldots\ldots\ldots(19).$$

26. When P is inside the magnetised body the integral of formula (14) is convergent, and so the formula may stand as the definition of the potential at P. The properties of V, thus defined, are most easily deduced from another expression, obtained by making a cavity round P and applying the theorem of § 17, formula (4), taking ϕ to be r^{-1} and writing A, B, C for ξ, η, ζ. The surface integral over the cavity has a zero limit, and so we get

$$V = \int_T (lA + mB + nC)\,r^{-1}dS - \int^T \left(\frac{\partial A}{\partial x} + \frac{\partial B}{\partial y} + \frac{\partial C}{\partial z}\right) r^{-1}d\tau \quad\ldots(20),$$

where the surface integral refers only to the boundary T of the body, and the cavity is now closed and finished with. This form of the magnetic potential exhibits it as equivalent to the *gravitation* potential of a volume distribution of density

$$\rho = -\frac{\partial A}{\partial x} - \frac{\partial B}{\partial y} - \frac{\partial C}{\partial z},$$

combined with a distribution of surface density $lA + mB + nC$ spread over the boundary of the body. (Of course formula (20) is equally true when P is outside the body.)

If we suppose that P is right inside the body, i.e. not on the boundary, there is no infinity in the subject of surface-integration; the volume-integral part of V has the properties of the gravitation potential studied in Section VI. Thus V has definite space differential coefficients obtained by differentiating under the sign of integration in formula (20) (not in formula (14)*); accordingly, since

$$\frac{\partial}{\partial \xi}(r^{-1}) = -\frac{\partial}{\partial x}(r^{-1}),$$

$$-\frac{\partial V}{\partial \xi} = \int_T (lA + mB + nC)\frac{\partial}{\partial x}(r^{-1})\,dS - \int^T \left(\frac{\partial A}{\partial x} + \frac{\partial B}{\partial y} + \frac{\partial C}{\partial z}\right)\frac{\partial}{\partial x}(r^{-1})\,d\tau,$$

* The theorem of § 21 does not apply to the integral of formula (14), since the infinity at P is of the order r^{-2}.

the volume integral having a perfectly definite value. Now we cut a cavity σ round P, and apply the theorem of § 17, formula (4), and we get

$$-\frac{\partial V}{\partial \xi} = -\operatorname*{Lim}_{\sigma \to 0} \int_\sigma (lA + mB + nC) \frac{\partial}{\partial x}(r^{-1})\, dS$$

$$+ \operatorname*{Lim}_{\sigma \to 0} \int_\sigma^T \left(A \frac{\partial}{\partial x} + B \frac{\partial}{\partial y} + C \frac{\partial}{\partial z}\right) \frac{\partial}{\partial x}(r^{-1})\, d\tau \quad \ldots(21).$$

These two limits combined give a value which is independent of the shape of σ, but the value of each limit taken separately depends on the shape of σ; the volume integral we see, by comparison with (15), to be the x component of the magnetic force in the cavity due to all the matter outside the cavity. So in general $-\frac{\partial V}{\partial \xi}$ is not the limit of the component of force in the cavity, but differs from it by an amount represented by the limit of the surface integral. If, however, we can choose a shape for the cavity which shall make the limit of the surface integral zero, the limit of the force component in the cavity will be accurately represented by $-\frac{\partial V}{\partial \xi}$; and this is effected by making the cavity a cylinder whose generators are parallel to the direction of the vector (A, B, C) at the point P, with flat ends perpendicular to the generators, all the linear dimensions of the cylinder tending to zero in such fashion that the linear dimensions of the ends tend to become vanishingly small compared with the length; this may, for brevity, be called a 'long' cylinder. The direction chosen for the generators ensures that the integral of $lA + mB + nC$ for the curved portion of the surface tends to zero, and the relative smallness of the flat ends makes the integral over these tend also to zero. The *definition* of the magnetic force (α, β, γ) at a point P in the body is 'the limit of the force in a cavity in the form of a long cylinder with generators parallel to the resultant intensity of magnetisation'; and this definition, in connexion with the present argument, justifies the statement that

$$\alpha = -\frac{\partial V}{\partial \xi}.$$

The *definition* of the induction (a, b, c) at a point P in the body is 'the limit of the force in a cavity in the form of a very flat circular cylinder with generators parallel to the resultant intensity of magnetisation,' where by a very flat cylinder is meant one whose linear dimensions tend to zero in such a way that the length tends to become vanishingly small in comparison with the linear dimensions of the plane ends. For such a cavity $lA+mB+nC$ tends to zero on the curved part of the surface, to $-I$ over one of the plane ends, and to $+I$ over the other, I being the resultant intensity of magnetisation; and each of these ends ultimately subtends a solid angle 2π at P. Thus the three surface integrals of which that in (21) is a type have for limits the components of force at a point between two infinite circular* parallel planes, the one covered with a uniform surface density I, the other with a uniform surface density $-I$, of matter that attracts according to the Newtonian law; this force is known to be $4\pi I$ perpendicular to the planes, and so its components are $4\pi A$, $4\pi B$, $4\pi C$. So, for the flat cavity, (21) yields the equality

$$a = -4\pi A + a \qquad \text{(22)}.$$

27. The integrals of formulae (18) representing vector potential are convergent for a point inside the body, and may therefore stand as the definition of the vector potential at such a point. If in the formula (4) of § 17 we put 0 for ξ, $-C$ for η, B for ζ, and r^{-1} for ϕ, we get

$$F=\int_T (nB-mC)\,r^{-1}\,dS-\int^T\left(\frac{\partial B}{\partial z}-\frac{\partial C}{\partial y}\right)r^{-1}d\tau \qquad \text{(23)},$$

the cavity of formula (4) being closed and finished with, and the surface integral over the cavity having a zero limit. This result exhibits F as the gravitation potential of a finite volume distribution combined with a surface distribution; it shews, therefore, that F has definite differential coefficients with respect to the coordinates of P.

From the two formulae analogous to (23),

$$\frac{\partial H}{\partial \eta}-\frac{\partial G}{\partial \zeta}=\int_T\left[(Cl-An)\frac{\partial}{\partial z}-(Am-Bl)\frac{\partial}{\partial y}\right]r^{-1}dS$$
$$+\int^T\left[\left(\frac{\partial A}{\partial y}-\frac{\partial B}{\partial x}\right)\frac{\partial}{\partial y}-\left(\frac{\partial C}{\partial x}-\frac{\partial A}{\partial z}\right)\frac{\partial}{\partial z}\right]r^{-1}d\tau \qquad \text{(24)}.$$

Cut a cavity σ round P and apply the theorem of formula (4), § 17, and the result is readily seen to be

$$\frac{\partial H}{\partial \eta}-\frac{\partial G}{\partial \zeta}=-\operatorname*{Lim}_{\sigma\to 0}\int_\sigma\left[(Cl-An)\frac{\partial}{\partial z}-(Am-Bl)\frac{\partial}{\partial y}\right]r^{-1}dS$$
$$+\operatorname*{Lim}_{\sigma\to 0}\int_\sigma^T\left[\left(C\frac{\partial}{\partial x}-A\frac{\partial}{\partial z}\right)\frac{\partial}{\partial z}-\left(A\frac{\partial}{\partial y}-B\frac{\partial}{\partial x}\right)\frac{\partial}{\partial y}\right]r^{-1}d\tau \qquad \text{(25)},$$

* The word 'circular' is introduced in order to exclude cases in which the resultant force at a point between the parallel planes is not normal to them. The circles are supposed to have a common axis, passing through P.

wherein the limits on the right-hand side together give a value independent of the shape of σ, though the value of each separately depends on the shape of σ. The volume integral is the same as

$$\int_\sigma^T \left(A \frac{\partial}{\partial x} + B \frac{\partial}{\partial y} + C \frac{\partial}{\partial z} \right) \frac{\partial}{\partial x} (r^{-1})\, d\tau$$

and accordingly represents the x component of force due to all the magnetisation outside the cavity; the surface integral, together with the corresponding surface integrals in the two other formulae analogous to (25), will tend to zero if $A/l = B/m = C/n$ over practically the whole surface of the cavity, and this is ultimately the case when the cavity is the flat cylinder used in defining the induction. For this shape of cavity (25) is equivalent to

$$\frac{\partial H}{\partial \eta} - \frac{\partial G}{\partial \zeta} = a \quad \ldots\ldots\ldots\ldots\ldots\ldots\ldots\ldots(26),$$

which, with the two other equalities of the same type, constitutes the vector relation

$$(a, b, c) = \text{curl}\,(F, G, H),$$

true now for points inside as well as for points outside the magnetised body.

It should be noticed that the definition of vector potential used in the present discussion is not that which is regarded as fundamental in the physical theory, though equivalent to it. The usual definition is contained in the relation (19) coupled with the relation

$$\frac{\partial F}{\partial \xi} + \frac{\partial G}{\partial \eta} + \frac{\partial H}{\partial \zeta} = 0.$$

It is easy to verify, on the lines of the present Article, that the vector defined by the relation (18), whether for internal or for external points, satisfies this further condition.

VIII. Surface Integrals.

28. When gravitating matter is distributed in a very thin layer, or when the surface of a body is charged with electricity, the corresponding potential and attraction at a point P are represented by surface integrals, a surface density σ taking the place of a volume density. The integrals are of the type $\int \sigma f dS$ where σ is usually free from such mathematical peculiarities as might raise doubts concerning the existence of the integrals, and f is a function having an infinity at the point P (ξ, η, ζ).

So long as P is not actually in the region of integration the integrals do not present any difficulties, and the formulae $X = \frac{\partial V}{\partial \xi}$, $\Delta V = 0$, are clearly valid.

When the point P is in the surface distribution we must cut a cavity round it of dimensions that tend to zero, and the question of convergence necessarily arises. We consider first the case in which the integration takes place over a portion of a *plane* surface.

The chief test of convergence is now as follows. *If within a circle of finite radius* (a), *having the point P as centre, the subject ϕ of integration is always less in absolute value than $Mr^{-\mu}$, where $\mu < 2$ and M is a definite constant, the integral $\int \phi dS$ is convergent.* To prove this we shall shew that, corresponding to any arbitrarily chosen small quantity σ, there can always be found a closed curve θ surrounding P such that all closed curves t surrounding P and lying wholly inside θ have the property that

$$\left| \int_t^\theta \phi dS \right| < \sigma.$$

Take for the curve θ the circle $r = \eta$, where $\eta < a$, and denote by ϵ the distance from P to the nearest point of the boundary t of the cavity; the cavity is of course entirely inside θ, but is otherwise unrestricted as to shape. Since the modulus of a sum is not greater than the sum of the moduli,

$$\left| \int_t^\theta \phi dS \right| \leqslant \int_t^\theta |\phi| \, dS,$$

$$\leqslant \int_\epsilon^\theta |\phi| \, dS, \quad < M \int_\epsilon^\theta r^{-\mu} dS,$$

$$< 2\pi M \int_\epsilon^\eta r^{1-\mu} dr, \qquad < \frac{2\pi M}{2-\mu} (\eta^{2-\mu} - \epsilon^{2-\mu}), \qquad < \frac{2\pi M}{2-\mu} \eta^{2-\mu}.$$

Hence by choosing η less than $\{(2-\mu)\sigma/2\pi M\}^{\frac{1}{2-\mu}}$ we get a curve θ satisfying the specified condition; the integral is accordingly convergent.

When the order of the infinity of ϕ is the same as that of r^{-2}, semi-convergence may appear.

29. Passing to the case in which the region of integration is a portion of a *curved* surface, we shall assume P to be a point at which there is a definite tangent plane and such that at all points of the region within a finite distance of P the principal curvatures are both finite. We need consider only the integral taken through a finite region not extending far from P, and in virtue of the finiteness of the curvatures at and near P we can always choose this region so that, if Q is any point of it and θ the inclination of the tangent plane at Q to the tangent plane at P, for all positions of Q in the region $\theta < \alpha$, where α is a definite acute angle. Let the projection of Q on the tangent plane at P be Q_0, let r, r_0 represent PQ, PQ_0 respectively, dS an element of area round Q, dS_0 the projection of dS on the tangent plane at P, B the boundary of the area of integration, and B_0 its projection on the tangent plane at P.

Since $dS = dS_0 \sec \theta$,

$$\int^{B} \phi \, dS = \int^{B_0} \phi \sec \theta \, dS_0,$$

the second integral being taken in the tangent plane at P.

If within the region of integration

$$|\phi| < Mr^{-\mu},$$

where M is a constant and $2 > \mu > 0$, then

$$|\phi \sec \theta| < Mr^{-\mu} \sec \theta$$
$$< Mr_0^{-\mu} (r_0/r)^{\mu} \sec \theta,$$

or, since $r_0 < r$, and $\sec \theta < \sec \alpha$,

$$|\phi \sec \theta| < M \sec \alpha \, r_0^{-\mu},$$

where $M \sec \alpha$ is finite since α is acute.

Hence $\int^{B_0} \phi \sec \theta \, dS_0$ is convergent, and therefore so also is $\int^{B} \phi \, dS$; thus the test of convergence is the same whether the surface of integration be plane or curved provided the curvatures be finite. The existence of a definite tangent plane at P is not a necessary feature in the proof, the essential thing is that there shall be a finite region round P for which $\theta < \alpha < \frac{1}{2}\pi$, θ being inclination to some fixed plane through P; for example the surface might be a cone and P its vertex. (Compare Poincaré, *Potentiel Newtonien*, § 33.)

30. Applying the test of the preceding Articles we see that at a point in a surface distribution of gravitating matter or electricity the potential is represented by a convergent integral, but the attraction components in the tangent plane are represented by integrals whose order renders semi-convergence possible. It is not difficult to shew, by a particular example, that semi-convergence does occur; for the attraction of a uniform plane elliptic disc (of eccentricity e and surface density σ) at a focus is $2\pi\sigma(1 - \sqrt{1 - e^2})/e$ if the cavity is circular, but is zero if the cavity is an ellipse similar and similarly situated to the edge of the disc, with the focus for centre of similitude; the verification of these statements, by using polar coordinates and integrating, is quite easy.

The component of attraction at P normal to the surface is represented by a convergent integral, but this quantity is the attraction *in a cavity*, though a vanishing one, and must be distinguished from the normal component of attraction at a point very close to the *unbroken* surface but not in it; it is, in electrical applications, the 'mechanical force per unit charge,' the quantity denoted by R_2 in Prof. Sir J. J. Thomson's *Elements of Electricity and Magnetism*, § 37, whereas the normal attraction at a point just not in the surface is the quantity there denoted by R.

31. The distinction drawn above, between the attraction at a point in the surface and that at a point just not in the surface, brings us to a question of a kind which, for lack of a fourth dimension, does not arise geometrically in the case of volume integrals, the question, namely, whether an integral $\int \phi dS$ tends to a definite limit if the point P, where ϕ has an infinity, is not originally in the surface, but approaches a point of the surface as a limiting position.

Let O be the point of the surface to which P gets continually nearer; it will be convenient to take O as origin of coordinates and the tangent plane at O as plane of z; we shall suppose that there is a limiting position of the line PO, as P moves up to coincidence with O, which makes with the plane of z a definite angle different from zero and so has definite direction cosines l_0, m_0, n_0, of which the last is numerically greater than zero. The length PO will be denoted by κ, and the coordinates of P by (ξ, η, ζ) or $(-l\kappa, -m\kappa, -n\kappa)$, while x, y, z represent the coordinates of a variable point Q on the surface; the subject of integration, $\phi(x, y, z, \xi, \eta, \zeta)$, may for brevity be represented by ϕ, while $\phi(x, y, z, 0, 0, 0)$, the value of ϕ when P is coincident with O, will be represented by ϕ_0. If integration be extended to a finite part of the surface round O, bounded by a closed curve B, the quantities to whose different meanings and possibly different values it is desired to draw attention are respectively

$$\int^B \phi_0 \, dS \text{ and } \operatorname*{Lim}_{\kappa \to 0} \int^B \phi \, dS.$$

The first thing to notice is that, while the integral of ϕ requires no cavity so long as κ is different from zero, which is the case at all stages in the passage to limit denoted by $\kappa \to 0$, the integral of ϕ_0 is only intelligible in terms of a cavity ϵ round the point O, though this cavity of course tends to vanishing. If, therefore, we set out to find the algebraic difference between the two quantities which form the subject of discussion (which may conveniently be denoted by D) we have

$$D \equiv \operatorname*{Lim}_{\kappa \to 0} \int^B \phi\, dS - \int^B \phi_0\, dS$$

$$= \operatorname*{Lim}_{\kappa \to 0} \int^B \phi\, dS - \operatorname*{Lim}_{\epsilon \to 0} \int_\epsilon^B \phi_0\, dS.$$

Since the term involving the limit for $\kappa \to 0$ in no way depends upon ϵ, and the term involving the limit for $\epsilon \to 0$ in no way depends upon κ, it is a matter of indifference in what order we suppose the passages to limit to be made; accordingly we are at liberty, if we please, to make first the passage to the limit for $\kappa \to 0$, so that at any stage short of the limits we shall think of κ as extremely small compared with the linear dimensions of the cavity ϵ. The difference, then, before passage to either limit, may be put in the form

$$\int^\epsilon \phi\, dS + \int_\epsilon^B \phi\, dS - \int_\epsilon^B \phi_0\, dS.$$

Now if we proceed first to the limit for $\kappa \to 0$, the points P and O at all stages of this passage are quite outside the region of integration of the last two integrals, and the functions ϕ and ϕ_0 are kept definitely removed from their infinite values; hence in the absence of peculiarities of ϕ other than that infinity at P which is the special subject of our investigation, we get the same limit for $\int_\epsilon^B \phi\, dS$ whether we first integrate and then make $\kappa \to 0$ or first make $\kappa \to 0$ and then integrate. In fact

$$\operatorname*{Lim}_{\kappa \to 0} \int_\epsilon^B \phi\, dS = \int_\epsilon^B \operatorname*{Lim}_{\kappa \to 0} \phi\, dS = \int_\epsilon^B \phi_0\, dS;$$

whence

$$D = \operatorname*{Lim}_{\epsilon \to 0} \left[\operatorname*{Lim}_{\kappa \to 0} \int^\epsilon \phi\, dS + \operatorname*{Lim}_{\kappa \to 0} \int_\epsilon^B \phi\, dS - \int_\epsilon^B \phi_0\, dS \right]$$

$$= \operatorname*{Lim}_{\epsilon \to 0} \operatorname*{Lim}_{\kappa \to 0} \int^\epsilon \phi\, dS \quad\ldots\ldots\ldots\ldots(27),$$

the notation implying that ϵ is kept constant while $\kappa \to 0$, thus yielding a limit which is a function of ϵ, and that *afterwards* the limit of this function of ϵ is taken for $\epsilon \to 0$.

Let us suppose ϕ to be of the form $(z-\zeta)^{\lambda}\, r^{-\mu}$ where r denotes PQ and λ and μ are positive, and let us proceed to make a closer examination of D for this particular case. We shall assume that the principal curvatures of the surface are finite at all points in a finite region round O, and that the cavity ϵ is determined by the intersection of the surface with the narrow cylinder $x^2+y^2=\epsilon^2$; and we picture to ourselves a small piece of the surface, which we may call the 'cap,' bounded by this curve whose projection on the plane of z is a circle of radius ϵ and centre O, and a point P at a distance κ from O which is extremely small compared with ϵ. To begin with, we observe that there is a finite constant a such that for all points Q in the cap $|z|<as^2$, where s stands for $\sqrt{x^2+y^2}$, the distance of Q from the axis of z; for, since the surface has finite curvature at all points of the cap, $|z|/s^2$ tends to a finite limit for any given azimuth as Q approaches O, and so is finite at all points of the cap; and the various values, being finite, have a finite superior limit a which is of the same order of magnitude as the greatest curvature of a normal section through O; thus the inequality is proved.

Consider now the curve of intersection of the surface with the cylinder $x^2+y^2=\kappa^2$; this divides the cap into two regions, an inner region whose linear dimensions are of the order of κ and therefore small in comparison with those of the cap, and an outer region comprising most of the cap, in which there is no point whose distance from O is of a higher order of smallness than κ. The integral whose limit is D can be regarded as the sum of two integrals, one over the inner region, one over the outer region, and these will be considered separately.

Taking first the outer region, and remembering the assumption $n_0 \neq 0$, we see that, while there may be points in this region for which $r<s$, there are no points in it for which s/r becomes infinite, so that there is a finite superior limit β for the values of s/r in the region; the finiteness of the curvature is a guarantee that there is a finite superior limit c (differing from unity by a quantity of the order of ϵ^2) to the secant of the inclination to the axis of z of the normal to the surface at Q, i.e. the ratio of dS to its projection dS_0 on the plane of z; and $|\zeta|/s$ is finite at all points of the region and has therefore a finite superior limit γ', so that $|\zeta|<\gamma' s$; as $|z|<as^2$, it follows that

$$|z-\zeta|\, s^{-1}<\gamma'+as<\gamma$$

where γ is a finite quantity.

Thus in the outer region

$$|\phi| = |(z-\zeta)^\lambda r^{-\mu}| < \gamma^\lambda s^\lambda \beta^\mu s^{-\mu},$$
$$dS < c\,dS_0,$$

and so

$$\left|\int \phi\, dS\right| < \gamma^\lambda \beta^\mu c \int s^{\lambda-\mu} dS_0,$$
$$< 2\pi\gamma^\lambda\beta^\mu c \int_\kappa^\epsilon s^{\lambda-\mu+1} ds,$$
$$< 2\pi\gamma^\lambda\beta^\mu c\,(\lambda-\mu+2)^{-1}\left[\epsilon^{\lambda-\mu+2}-\kappa^{\lambda-\mu+2}\right].$$

The limit of this for $\kappa \to 0$ and $\epsilon \to 0$ is zero provided $\mu - \lambda < 2$.

Taking next the inner region, we know that in it $|z| < as^2 < a\kappa^2$, while ζ is of the same order of smallness as κ, so that $|z-\zeta|$ is of the same order as κ, and there must be an inequality $|z-\zeta| < g\kappa$ where g is a definite constant. Clearly there is a superior limit to the secant of the angle between the normal at Q and the line PQ, so that there is an inequality $dS < hr^2 d\omega$, where $d\omega$ represents an element of solid angle at P; and r always bears to κ a finite ratio, so that there is a double inequality $p\kappa > r > q\kappa$, where p and q are constants. Hence in the inner region

$$\int (z-\zeta)^\lambda r^{-\mu} dS < g^\lambda q^{-\mu} p^2 h \kappa^{\lambda-\mu+2} \int d\omega,$$

and the limit of this, for $\kappa \to 0$, is zero provided $\mu - \lambda < 2$.

Thus $D = 0$ for $\phi = (z-\zeta)^\lambda r^{-\mu}$ subject to $\mu - \lambda < 2$, and it is an immediate inference that $D = 0$ for $\phi = \sigma(z-\zeta)^\lambda r^{-\mu}$ subject to the same condition, if σ is a function of x, y, z which is finite throughout the region of integration. For this type of integral the condition for the vanishing of D is not the same as the condition for the convergence of the integral of ϕ_0, for, in the neighbourhood of O, z becomes small of the order of r_0^2, so that the condition of convergence of ϕ_0 is

$$\mu - 2\lambda < 2.$$

Powers of $x-\xi$, $y-\eta$ might appear in ϕ; they would count as the corresponding powers of r in applying the test just proved, though of course they would not be equivalent to powers of r if one were examining a case where something analogous to semi-convergence seemed probable.

For the potential integral $\lambda = 0$, $\mu = 1$, and therefore $D = 0$. Thus the potential at O is the same as the limit of the potential at P as P approaches O, so that V is not discontinuous at points on the surface.

32. For the attraction components $\mu - \lambda = 2$, and the test of the previous Article is not applicable so that a special investigation is required. Let us consider first a tangential component, say that whose subject of integration is $\sigma(x-\xi)r^{-3}$.

The corresponding integral of ϕ_0 is semi-convergent, and it may appear useless to investigate the difference D between the unknown limit of the integral of ϕ and the quantity of uncertain value which is the integral of ϕ_0. But the integral of ϕ is quite independent of the shape of the cavity, in fact it does not require any cavity, so we are as free as in the case of absolute convergence to give to the cavity any shape we please, so long as we attach to the integral of ϕ_0 the value associated with that particular shape; thus the semi-convergence does not introduce any uncertainty into the meaning of D.

Let us take the same cavity as in the preceding Article, using the same notation and applying whatever parts of the reasoning remain valid for the changed form of ϕ. And let us consider what error is introduced into the subject of integration if we replace σ by σ_0, its value at O, dS by its projection dS_0 on the plane of z, and r by r' the distance from P to the projection of Q on the plane of z. The error due to the change in σ corresponds to the omission of a factor which differs from unity by a quantity of the order of smallness of s, if we assume the function σ to have no troublesome peculiarities* at O; the error due to the change in dS corresponds to the omission of a factor differing from unity by a quantity of the order of s^2; and, since $|r-r'|<|z|<as^2$, the error due to the change in r corresponds to the omission of a factor differing from unity by a quantity of the order s^2r^{-1}. So the most important terms representing error in the integral over the cap are of the order of the integrals over the cap of $(x-\xi)r^{-3}s$ and $(x-\xi)r^{-4}s^2$; for these error integrals $\lambda=0$, $\mu=1$, s in the numerator counting as equivalent to r since there is a finite superior limit to sr^{-1}, and so, by the previous Article, the error has a zero limit for $\kappa \to 0$ and $\epsilon \to 0$.

Hence, for the x component of attraction,

$$D = \operatorname*{Lim}_{\epsilon \to 0} \operatorname*{Lim}_{\kappa \to 0} \int^{\epsilon} \sigma_0 (x-\xi) r'^{-3} dS_0,$$

where it is clear that the integral is now taken over a circular area of

* If further precision be desired, we may assume $|\sigma-\sigma_0|<Ms^m$, where M is finite and m positive. The error corresponding to this is less than the integral of $Ms^m(x-\xi)r^{-3}$, for which $\lambda=0$, $\mu=2-m$, and the limit of the error is zero. In the text m is taken to be unity.

radius ϵ in the plane of z, and represents the component of attraction at P of a circular disc of uniform surface density σ_0. Now, before passage to the limit, P is not in such a position of symmetry that the x attraction component must vanish, but if we describe the reflexion with respect to the plane $x = \xi$ of the lesser of the two arcs into which this plane divides the circumference of the disc, we obtain a division of the disc into two areas the greater of which, on account of the symmetry of the position of P with respect to it, contributes nothing to the attraction component. The component is therefore that due to the crescent-shaped smaller area, whose mass is very nearly $4\xi\epsilon\sigma_0$ and whose nearest point is at a distance from P comparable with ϵ; thus the component of attraction is of the same order of magnitude as $\sigma_0\xi\epsilon^{-1}$, which has a zero limit if we make $\kappa \to 0$ (i.e. $\xi \to 0$) before $\epsilon \to 0$. Thus D is zero, so that the x component of attraction of the whole surface at P tends to a limit, as P approaches O, equal to the corresponding component of attraction at O reckoned for a vanishing circular cavity with O as centre; the limit is the same from whichever side of the surface P moves up to O.

33. In the case of the normal attraction component Z, the subject of integration is $\sigma(z-\zeta)r^{-3}$, and the corresponding component at O is represented by a convergent integral. Thus

$$D = \operatorname*{Lim}_{\epsilon \to 0} \operatorname*{Lim}_{\kappa \to 0} \int^{\epsilon} \sigma(z-\zeta)\, r^{-3} dS.$$

The most important terms of the error introduced into the integral of this formula by putting σ_0 for σ, r' for r, and dS_0 for dS, are of the same order of magnitude as

$$\int^{\epsilon} \sigma_0(z-\zeta)\, sr^{-3} dS^{*} \quad \text{or} \quad \int^{\epsilon} \sigma_0(z-\zeta)\, s^2 r^{-4} dS;$$

for both of these $\lambda = 1$, $\mu = 2$, and therefore, by § 31, the error has a zero limit. Accordingly D is the limit of

$$\int^{\epsilon} \sigma_0 z r'^{-3} dS_0 - \int^{\epsilon} \sigma_0 \zeta r'^{-3} dS_0;$$

and in the former of these we notice that z is of the order of smallness of s^2, and therefore the integral of the same order as $\int^{\epsilon} \sigma_0 s^2 r'^{-3} dS_0$, which has $\lambda = 0$, $\mu = 1$, and therefore the limit zero. Thus

$$D = -\operatorname*{Lim}_{\epsilon \to 0} \operatorname*{Lim}_{\kappa \to 0} \sigma_0 \int^{\epsilon} \zeta r'^{-3} dS_0,$$

* If we make the same assumption with regard to σ as in the footnote of § 32 the index of s will be m in this integral.

the integral being now taken over the plane circular area bounded by $r'=\epsilon$, $z=0$. Now if $d\omega$ represent the solid angle subtended by dS_0 at P, $|\zeta| r'^{-3} dS_0 = d\omega$, and the integral is $\pm \int d\omega$, the sign being positive if ζ is positive, negative if ζ is negative. If we make $\kappa \to 0$ before $\epsilon \to 0$, clearly the limit of $\int d\omega$ is 2π, and so we get

$$D = \mp 2\pi\sigma_0,$$

the upper sign corresponding to ζ positive. So the limit of Z differs from the value of Z at O by $2\pi\sigma_0$, the excess of the former over the latter corresponding to an attraction $2\pi\sigma_0$ *towards* the surface; the difference between the limits of Z as P approaches O from different sides of the surface is $4\pi\sigma_0$, and the arithmetic mean of these limits is the value of Z at O.

34. The potential at P of a double sheet, or normally magnetised shell, of strength μ' at the point Q, is given by

$$V = \int \mu' r^{-2} \cos\psi \, dS,$$

where ψ is the angle between QP and the normal at Q drawn in the sense for which μ' is reckoned positive. The error introduced into the subject of integration by taking ψ at points near O to mean the angle between QP and the axis of z, and so replacing $\cos\psi$ by $-(z-\zeta)r^{-1}$, corresponds to dropping a factor which differs from unity by a quantity of the order of s^2, and the integral of this error taken over the cap is one for which, in the notation of § 31, $\lambda = 1$, $\mu = 1$, and therefore has a zero limit. Hence the potential of a double sheet has, for purpose of finding D, the same form as the integral investigated in the preceding Article; and the limit of V as P approaches O from the positive side of the sheet exceeds by $4\pi\mu_0'$ the limit as P approaches O from the negative side.

35. The potential of a surface distribution of gravitating matter, whose surface density is free from such peculiarities as would render invalid the properties already established, has in a certain sense a space differential coefficient in any direction at any point O of the surface; this is not a differential coefficient as generally defined, since it is a limit which has different values according as the consecutive point P approaches O from one side of the surface or from the other. It is to be noticed that the existence of a differential coefficient cannot be inferred from the physical property that force equals gradient of potential, since O is a point not in free space, but in the gravitating matter.

The theorem is that

$$\lim_{OP\to 0}\{(V_O - V_P)/OP\} = \lim_{OP\to 0} F_P,$$

where F_P is the component of the force at P resolved along the tangent to the path by which P approaches O.

To prove this we must shew that if η be any arbitrary small quantity we can always choose a point K on the curve by which P approaches O such that for all positions of P on the curve between K and O

$$\left|\frac{V_O - V_P}{OP} - F_O\right| < \eta,$$

where F_O represents $\lim_{OP\to 0} F_P$.

Let us regard η as the sum of three arbitrary parts η_1, η_2, and η_3. We take a point J on the curve between P and O, and notice that

$$\frac{V_O - V_P}{OP} - F_O = \frac{V_O - V_J}{OP} + \frac{V_J - V_P}{JP}\cdot\frac{JP}{OP} - F_O\,;$$

and, remembering that

$$V_J - V_P = \int_P^J F ds,$$

where F is the tangential force and ds an element of the curve, we apply the first theorem of mean value and so get

$$V_J - V_P = JP\,.\,F_Q,$$

where Q is some point on the curve between J and P.

Thus

$$\frac{V_O - V_P}{OP} - F_O = \frac{V_O - V_J}{OP} + \frac{JP}{OP}\cdot F_Q - F_O.$$

Since F is definite at all points between O and P, and has a definite limit for a point tending to coincidence with O, there is a definite superior limit to the absolute value of F for the points of the curve lying between O and any definite point K; we call this superior limit M.

Now we choose K so near to O that, for all points P between K and O, $|F_P - F_O| < \eta_1$ and therefore also $|F_Q - F_O| < \eta_1$; this we can do because F_P has the limit F_O.

We next take P anywhere on OK, and P having been chosen, we can choose a point L_2 so that for all points J between L_2 and O, $|V_O - V_J| < OP\,.\,\eta_2$, this being possible because V_J has the limit V_O.

And a point L_3 can be chosen so that for all points J between L_3 and O

$$\frac{JP}{OP} = 1 + \epsilon$$

where $|\epsilon| < \eta_3/M$. We now take J to be between O and the nearer of the points L_2, L_3.

Thus

$$\frac{V_O - V_J}{OP} + \frac{JP}{OP} F_Q - F_O = \frac{V_O - V_J}{OP} + (1+\epsilon) F_Q - F_O$$

$$= \left[\frac{V_O - V_J}{OP}\right] + [F_Q - F_O] + \left[\epsilon M \cdot \frac{F_Q}{M}\right],$$

where we notice that $|F_Q| < M$. The modulus of the first expression in square brackets is less than η_2, that of the second is less than η_1, that of the third is less than η_3; hence the modulus of the sum of the three expressions is less than $\eta_1 + \eta_2 + \eta_3$ or η. Thus we have been able to choose K so that for all points P on the curve between O and K

$$\left|\frac{V_O - V_P}{OP} - F_O\right| < \eta,$$

which establishes the theorem.

IX. Volume Integrals through regions that extend to infinity.

36. The integrals to which we have so far been devoting most attention are those whose peculiarity consists in the subject of integration becoming infinite at a point in the range. Another kind of integral requiring special study occurs frequently in mathematical physics, namely, a volume integral taken through a region which extends to infinity.

By the integral $\int f d\tau$ taken through all space outside certain finite closed surfaces S_1, S_2, etc. is meant the limit of the integral taken through a region bounded internally by S_1, S_2, etc., and externally by a surface B, as the linear dimensions of B and the distances of all its points from the inner boundaries become indefinitely great, provided such a limit exists and is independent of the shape of B. When the limit exists and is independent of the shape of B, the integral is said to be convergent; if the limit has a finite value which is not independent of the shape of B, the integral is said to be semi-convergent.

The following is the chief test of convergence. *If we measure r from some fixed origin, and if f is such that, for all values of r greater than a definite length a, f is less in absolute value than $Mr^{-\mu}$, where M is a constant and $\mu > 3$, the integral is convergent.* We shall prove this by shewing that, corresponding to any arbitrarily chosen small quantity σ, there can always be found a closed surface θ surrounding O and all the surfaces S_1, S_2, etc., such that all closed surfaces t surrounding θ have the property that

$$\left| \int_\theta^t f\, d\tau \right| < \sigma.$$

Take for the surface θ a sphere $r = \eta$ large enough to surround the sphere $r = a$ and all the inner boundaries of the region; and let ω be the distance from O to the furthest point of the outer boundary t. Then

$$\left| \int_\theta^t f\, d\tau \right| \leqslant \int_\theta^t |f|\, d\tau, \leqslant \int_\theta^\omega |f|\, d\tau,$$

the upper limit in the last integral being the sphere $r = \omega$. Thus

$$\left| \int_\theta^t f\, d\tau \right| < M \int_\theta^\omega r^{-\mu}\, d\tau, < 4\pi M \int_\eta^\omega r^{2-\mu}\, dr,$$

$$< \frac{4\pi M}{\mu - 3}\left(\eta^{-(\mu-3)} - \omega^{-(\mu-3)}\right), \text{ a positive quantity,}$$

$$< \frac{4\pi M}{\mu - 3}\eta^{-(\mu-3)}.$$

Hence by choosing η greater than $\{4\pi M/(\mu - 3)\,\sigma\}^{\frac{1}{\mu-3}}$ we get a surface θ satisfying the specified condition; the integral is accordingly convergent. It will be noticed that there is no restriction on the shape of the outer boundary t.

Generally speaking, if f is zero at infinity of an order higher than r^{-3}, the integral is convergent; if the zero is just of the order r^{-3}, the integral may be semi-convergent or divergent.

37. When Green's theorem and allied theorems are applied to volume integrals of this type, the outer boundary which tends to become infinitely large must not be left out of account, and so we have limits of surface integrals which are spoken of as integrals over the surface infinity. If the subject ψ of integration is, for values of r greater than a finite length a, less in absolute value than $Mr^{-\mu}$, where M is finite and $\mu > 2$, then $|\int \psi\, dS|$ taken over the sphere $r = \omega$ is less than $M\omega^{2-\mu}\int\int \sin\theta\, d\theta\, d\phi$, which has the limit zero for $\omega \to \infty$.

Thus the surface integral vanishes if ψ is zero at infinity of a higher order than r^{-2}. If ψ is zero of the order r^{-2} the limit of the surface integral may be different for different shapes of B; if this is the case there is of course corresponding semi-convergence of one of the volume integrals, and special investigation is required.

38. The differentiation with respect to a parameter of a volume integral through a region extending to infinity, involving as it does two distinct passages to limits, requires special consideration. Let us consider the case in which the parameter ξ affects the subject of integration, but does not affect the specification of the inner boundaries S_1, S_2, etc. Let us suppose that $\partial f/\partial\xi$ (or f') exists and is uniformly continuous through all finite portions of the region of integration for all values of ξ considered, and that the integral of f is convergent; and further that, for all values of ξ considered and for all values of r greater than a, there is an inequality $|f'| < Mr^{-\mu}$, where $\mu > 3$, and M, μ, and a are constants whose values do not depend on the value of ξ*. Take the outer boundary to be the sphere $r = \omega$; then

$$\frac{\partial}{\partial\xi}\int f\,d\tau - \int f'\,d\tau$$

$$= \lim_{\Delta\xi\to 0}\ \lim_{\omega\to\infty}\left[\frac{1}{\Delta\xi}\int^{\omega}\{f(\xi+\Delta\xi) - f(\xi)\}\,d\tau - \int^{\omega} f'(\xi)\,d\tau\right] \quad \ldots(28),$$

and this, by the theorem of mean value,

$$= \lim_{\Delta\xi\to 0}\ \lim_{\omega\to\infty}\int^{\omega}\epsilon\,d\tau,$$

where $\epsilon \equiv f'(\xi + \theta\Delta\xi) - f'(\xi)$ and $1 > \theta > 0$, and the notation implies that first $\omega \to \infty$ and afterwards $\Delta\xi \to 0$. If we can shew that the subject of this double limit can, by first making $\omega \to \infty$, and afterwards taking $\Delta\xi$ sufficiently small, be made less than any arbitrarily assigned small quantity σ, clearly the double limit will be zero.

Now since f' satisfies the conditions of the theorem of § 36, and moreover in such a way that M, μ, and a are independent of ξ, it is clear by the reasoning of that Article that, for all values of ξ considered and therefore in particular for all possible values of $\xi + \theta\Delta\xi$, we can choose a definite length η such that for all values of ω greater than η

$$\left|\int_{\eta}^{\omega} f'(\xi + \theta\Delta\xi)\,d\tau\right| \text{ and } \left|\int_{\eta}^{\omega} f'(\xi)\,d\tau\right|$$

* We can get greater generality by simply requiring that the integral of f' shall be *uniformly* convergent for the contemplated range of values of ξ; but it seems better not to introduce into the text the idea of uniform convergence, especially as there are additional difficulties in the proof.

are both less than any assigned small quantity, which we shall take to be $\frac{1}{3}\sigma$. Hence

$$\lim_{\omega\to\infty}\left|\int_\eta^\omega f'(\xi+\theta\,\triangle\,\xi)\,d\tau\right|<\tfrac{1}{3}\sigma$$

and
$$\lim_{\omega\to\infty}\left|\int_\eta^\omega f'(\xi)\,d\tau\right|<\tfrac{1}{3}\sigma.$$

And η being chosen and therefore finite, however large, the uniform continuity of f' ensures our being able to choose a value of $\triangle\xi$ such that for it and for all smaller values $|\epsilon|$ is less than an arbitrary small quantity; this small quantity we choose to be $\frac{1}{3}\sigma T^{-1}$, where T is the finite volume $\int^\eta d\tau$. This makes $\left|\int^\eta \epsilon\,d\tau\right|<\frac{1}{3}\sigma$. Thus

$$\lim_{\omega\to\infty}\left|\int^\omega \epsilon\,d\tau\right|=\lim_{\omega\to\infty}\left|\int_\eta^\omega f'(\xi+\theta\,\triangle\,\xi)\,d\tau-\int_\eta^\omega f'(\xi)\,d\tau+\int^\eta \epsilon\,d\tau\right|<\sigma.$$

Hence the double limit on the right-hand side of (28) is zero, and therefore the differentiation of the integral of f is effected by the rule of differentiating under the sign of integration.

Differentiation with respect to a parameter which affects only the specification of one of the inner boundaries, say S_1, clearly gives rise merely to a surface integral over S_1.

39. Volume integrals through regions extending to infinity occur in electrical theory as expressions for electrostatic and for electrodynamic energy, and in other ways. They occur in the theory of gravitation and electrostatic potential in proofs of the important 'theorems of uniqueness.' They occur in Hydrodynamics as representing kinetic energy, and 'impulse.' Differentiation of such integrals is employed in the dynamical theory of solid bodies moving through an infinitely extended liquid. In every such application of these integrals it is necessary to make sure that there is such convergence as will render the formulae valid.

X. Gauss's Theorem in the Theory of Attractions*.

40. In a memoir on attractions and repulsions according to the law of the inverse square† Gauss enunciates the theorem that the surface integral of normal force taken over a closed surface is equal to

* This section is a reprint, with slight modifications, of a paper published in the *Proceedings of the London Mathematical Society*, Series 2, Vol. VIII. p. 200.

† C. F. Gauss, *Ges. Werke*, Bd. v., s. 224.

$4\pi M + 2\pi M'$, where M is the total mass of all the matter which is surrounded by the surface and M' the total mass of all the matter which lies as a surface distribution in the surface considered.

One way of proving this theorem is to deduce it from Poisson's equation (§ 23) by a volume integration, but a direct proof from first principles is preferable.

The proof given by Gauss and reproduced in most text-books implicitly involves the view that the attracting or repelling substance (whether matter or electricity) is made up of discrete particles of dimensions which are either absolutely zero or negligibly small in comparison with other distances in the contemplated configuration. It is shewn that the contribution of a particle of mass m to the surface integral of normal force is zero if the particle is definitely outside the space enclosed by the surface, and $4\pi m$ if the particle is definitely inside that space.

A particle of absolutely zero dimensions is necessarily in the enclosed region, outside the enclosed region, or in the surface which constitutes the boundary. In the last case it is easy to shew that the contribution of the particle to the surface integral of normal force is generally $2\pi m$, but may have some other value if the particle is at a singular point of the surface. There is therefore no difficulty in verifying, for point particles, the complete theorem as stated originally by Gauss, namely so as to include particles of no dimensions situated in the boundary.

When the particles considered have size, the ordinary argument applies, with as much precision as there is in our knowledge of the truth of the Newtonian Law, to such as are at a distance from the boundary great in comparison with their own linear dimensions. But in the case of particles, some of whose points are at a distance from the boundary which is not great in comparison with their linear dimensions, two difficulties arise. In the first place the law of the inverse square may fail adequately to represent the field of force at such close proximity to the particle; and, in the second place, in the absence of information as to the size and structure of a particle, it may be uncertain whether the particle is or is not wholly on one side of the boundary surface, and, if it is cut by the surface, how its contribution to the integral is to be reckoned.

The mathematical theory of attractions, however it may appear fundamentally and originally to have treated of particles, has by modern convention been in great measure transferred to another field of investigation; and the most familiar propositions of the theory are enunciated as applying to an ideal attracting substance which is not made up of discrete particles, i.e. is not of molecular structure, but is a continuum. The transition to such a substance from the particles originally discussed is made by dividing space into volume elements, and treating the continuous matter in a volume element as a particle. Clearly this is justifiable so long as it is recognized that the matter in a volume element constitutes a particle whose dimensions are not zero, and which therefore only comes under the elementary reasoning which leads to the Gauss theorem when it is at a distance from the boundary great compared with its linear dimensions.

Thus the transition, usually assumed without discussion, from the Gauss theorem for particles to the corresponding theorem for continuous matter is quite safe provided the surface S over which the integral is taken does not cut through the continuous matter. But the transition is not obviously safe, and requires special proof, for a surface that intersects the matter; for in this case some of the volume elements which are treated as particles must necessarily be actually in contact with the surface of integration.

41. In order to see what additional proof is required in this case, let us draw two surfaces parallel to the surface S at a small distance ϵ from it, the one S_1 inside it, the other S_0 outside it.

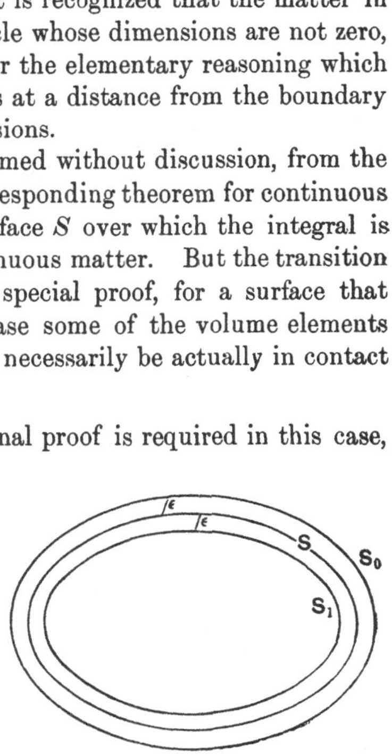

The normal force N at any point of the surface S may be regarded as made up of three parts, namely: (i) N_1, the part due to all the matter inside S_1, (ii) N_0, the part due to all the matter outside S_0, (iii) N', the part due to all the matter in the space between S_1 and S_0.

For any selected value of ϵ, the volume elements which take the place of particles can always be chosen so small that their linear dimensions are as small as we please in comparison with ϵ. Hence the ordinary elementary reasoning is valid for all the matter M_1 inside S_1, and for all the matter M_0 outside S_0. Hence, for the surface S,

$$\int N_1 dS = 4\pi M_1, \qquad \int N_0 dS = 0.$$

Thus $$\int N dS = \int N_1 dS + \int N_0 dS + \int N' dS = 4\pi M_1 + \int N' dS.$$

If now we pass to the limit, for $\epsilon \to 0$, it is obvious that Lim $M_1 = M$, where M is the mass of all the matter inside S*. Hence

$$\int N dS = 4\pi M + \lim_{\epsilon \to 0} \int N' dS;$$

so the Gauss theorem is true if, and only if,

$$\lim_{\epsilon \to 0} \int N' dS = 0.$$

42. A class of cases in which this limit is zero can easily be specified. For if, for any selected value of ϵ, there is a maximum value or superior limit to the values of $|N'|$ at points on S, say n, then

$$\left| \int N' dS \right| < nS,$$

where S is the complete area of the closed surface. And if, further,

$$\lim_{\epsilon \to 0} n = 0,$$

then clearly

$$\lim_{\epsilon \to 0} \int N' dS = 0.$$

We shall see that these conditions are satisfied in ordinary cases of matter so distributed that the volume-density is everywhere finite.

Volume-Density. If the distribution of continuous matter between S_1 and S_0 has everywhere finite volume-density, we know (§ 23) that the force-intensity at every point in that region is definite, and so N' is definite at every point of S. Consequently the superior limit or maximum value n exists. It remains to ascertain whether $n \to 0$ as $\epsilon \to 0$. The first step towards this is to shew that in all ordinary cases $N' \to 0$ as $\epsilon \to 0$. When this has been established, if n is a maximum value of $|N'|$, the convergence of all values of N' to zero necessarily involves also the convergence of n to zero. But if n is a superior limit to $|N'|$ without being a maximum value, we can be sure of the convergence of n to zero only if the convergence of N' to zero is uniform for the values of N' corresponding to all points of the surface S.

43. If we assume that the part of the material distribution contained between S_1 and S_0 consists only of finite volume-density which,

* M does not include matter distributed with finite surface-density *in* the surface S.

though not necessarily continuous, is a function of position free from other analytical peculiarities, then singularities in the function N' can only arise from peculiarities in the shape of the surface S and in the direction of the normal along which N' is the component of force. At an ordinary conical point, or at a point on the intersection of two sheets of the surface S, the direction of the normal is indeterminate and so also is the value of N'; nevertheless there are definite directions to which the normal tends as to a limit, and corresponding values which are limits of the function N' without being values of the function. One of the limits of N' at such a point might be the superior limit n and yet not be a maximum value. This sort of case must not be omitted from the present discussion, for one of the closed surfaces most frequently employed in applications of the Gauss theorem is a cylinder with flat ends, i.e. a surface of three sheets with two nodal lines.

The study of the influence of peculiarities of the surface S can be avoided if we express our test in terms of F', the resultant force-intensity due to the matter between S_0 and S_1, instead of N' its normal component. For $N' \leqslant F'$, and, if f be the maximum value or superior limit of F',

$$\left| \int N' dS \right| < fS,$$

and the Gauss theorem holds provided $f \to 0$ as $\epsilon \to 0$.

For the simple kinds of material distribution contemplated, F' is definite and continuous everywhere between S_1 and S_0, so there is no possibility of f being a superior limit without being a maximum value. And therefore, if we can shew that $F' \to 0$ for every point on S, we are thereby assured that $f \to 0$. The convergence of F' to zero is proved by shewing that the component G' of F' in any arbitrarily selected direction tends to zero.

44. We aim, then, at shewing that the component in any direction of the intensity of force at a point, due to a distribution of given volume-density contained between two parallel surfaces, tends to zero as the surfaces tend to coincidence. This result would be obvious if the point were definitely outside the attracting distribution. But it is not obvious in the present instance since the point is inside the distribution, nor would it be obvious if the point, though outside the distribution, were at a distance from its boundary which tended to vanishing.

Let P be the point at which F' is estimated. With P as centre, describe a sphere of radius θ. Then G' at P is the sum of two terms, G_1' due to that part of the matter between S_1 and S_0 which is exterior to the sphere, and G_2' due to that part of the matter between S_1 and S_0 which is interior to the sphere.

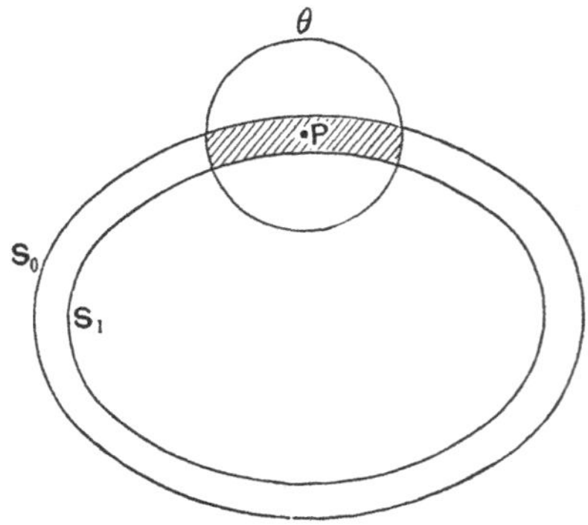

Now in consequence of the convergence of the integral representing the component of force in any direction due to a volume distribution, it is always possible to choose θ so small that G_2' shall be less than any assigned small quantity $\frac{1}{2}\omega$. And when θ has been selected we notice that P is definitely outside the matter which gives rise to G_1', so that G_1' tends to zero as the quantity of matter tends to zero. Thus we can always choose ϵ so small that G_1' shall be less than $\frac{1}{2}\omega$. Consequently we have been able to choose ϵ so small that

$$G_1' + G_2' = G' < \omega.$$

Thus
$$\lim_{\epsilon \to 0} G' = 0.$$

This is all we require for a proof of the Gauss theorem in the case in which there is no distribution of matter between S_0 and S_1 save such as has finite volume-density. We have seen that, since $G' \to 0$,

$$f \to 0 \quad \text{and} \quad \int N' \, dS \to 0.$$

So we conclude that when the surface S cuts only through matter of finite volume-density the Gauss theorem is valid.

45. *Surface-Density.* Let us now consider the case in which the distribution of matter between S_1 and S_0 includes a surface distribution, say on a surface Ω, of finite surface-density. In general the surface Ω will cut the surface S in a curve. A particular case, requiring special consideration, is that in which the surfaces S and Ω coincide over a definite area.

At the outset it is important to remember that the force-intensity at a point of a surface distribution such as Ω is represented by a semi-convergent integral (§ 30) and is therefore not completely defined.

By a suitable convention the definition may be completed, and in such a way that the force-intensity is finite. Adopting some such convention, we have N' finite at points on the curve of intersection of Ω and S; we have already seen that N' is finite at all other points of S, so that there is a finite superior limit or maximum value n of the values of $|N'|$ for all the points of S.

46. It is to be remarked that the particular convention adopted in order to give definiteness to N' does not generally affect the value of $\int N'dS$. If the surface Ω cuts the surface S in a curve, then the elements of area associated with dubious values of N' in the surface integral form a narrow strip on S along the curve of intersection of the two surfaces, and in the process of integration the area of this strip tends to zero, making its contribution to the surface integral also zero. On the other hand if the surface Ω coincides, over an area A, with the surface S, N' at points of A is the component of force-intensity normal to the former surface as well as to the latter. Now all the indeterminateness or semi-convergence of force due to a surface distribution is in the tangential component; the normal component is definite, and so there is no indeterminateness in the value of $\int N'dS$.

Incidentally we recall the fact (§ 33) that the normal force-intensity at a point in a surface containing surface-density σ differs from the limit of the normal force for a point approaching the surface but not in it by $2\pi\sigma$. Consequently the surface integral $\int N'dS$ is less when the surface of integration coincides with the surface on which the surface-density resides than if the former surface were just outside the latter by the amount $2\pi\int\sigma dS$ taken over the common area A.

47. In the more general case, in which Ω intersects S, having seen that $\int N'dS$ is definite we must examine what is the limit of its value as $\epsilon \to 0$. This we do by considering the component G' in any direction of the force-intensity at a point P in the curve of intersection of Ω and S, due to that part of the surface Ω which is intercepted between S_1 and S_0, and by investigating whether, as $\epsilon \to 0$, G' tends to infinity or to a finite or zero limit.

The problem is strictly analogous to that already discussed for volume-density, being, namely, that of examining the limit of a component of attraction at any point in a distribution of finite surface-density, as the distribution is diminished to zero in one of its dimensions without change of surface-density. The result is, however, quite different from that obtained for volume-density.

Let us begin by attempting to apply the method already used for volume-density, with such changes as are demanded by the changed circumstances. The attempt will be unsuccessful, but a study of the reason for its failure will help to make clear the nature of the difficulty which has to be faced. P is a point in the strip of breadth $2\epsilon \operatorname{cosec} \chi$ cut by S_0 and S_1 on the surface Ω, where χ is the angle of intersection of S and Ω. Suppose the ambiguity arising out of the semi-convergence of the surface integral representing G' to have been removed by selecting some special shape for the vanishing cavity round P. Describe round P, in the surface Ω, a curve θ of the special shape selected. The component of force at P due to the whole strip of Ω contained between S_1 and S_0 consists of two parts, namely, G_1' due to the part of the strip outside θ, and G_2' due to the part of the strip inside θ. On account of the convergence of the integral representing force due to a surface distribution, it is possible to choose the dimensions of θ so small that $|G_2'|$ is less than any selected small quantity $\frac{1}{2}\omega$. And when θ is chosen and fixed, P is definitely outside the distribution that gives rise to G_1', so clearly G_1' tends to zero as the distribution that gives rise to it tends to vanishing. Accordingly we can choose ϵ so small that $|G_1'| < \frac{1}{2}\omega$. Hence we have been able to choose ϵ so small that $G_1' + G_2' = G' < \omega$. Hence, apparently,

$$\operatorname*{Lim}_{\epsilon \to 0} G' = 0.$$

48. This reasoning, however, is not sound, and the result obtained is false. Semi-convergence is equivalent to convergence associated with a vanishing cavity about P of a definite shape, and implies that the limit of the attraction component (defined by means of such a vanishing cavity) of a portion of the distribution enclosed by a curve θ of the same shape about P is zero, as the linear dimensions of θ tend to zero. Now when the dimensions of θ have been selected and fixed in the above argument, the subsequent selection of ϵ may give us a strip (as in the diagram of § 44) which is narrower than θ, so that the whole of θ is not occupied by matter. Thus the actual matter within θ is bounded, not by θ, but by a curve of different shape, and the reasons for regarding the corresponding attraction at P as less than $\frac{1}{2}\omega$ cease to be applicable.

If we had absolute convergence instead of semi-convergence it would be quite another matter. For then we should not be tied to any particular shape of cavity or of boundary, and the fact of the boundary's becoming something else instead of θ would not invalidate the assumed inequality. We should, in fact, be entitled then to say that the attraction at P due to any area surrounding P and lying entirely within θ is less than $\frac{1}{2}\omega$. Now the component of force normal to Ω is represented by an absolute convergent integral, and so its limit is zero as the breadth of the strip tends to vanishing. But it is not so with the component in any other direction.

Consider, for example, a tangential component of force at P. For different shapes of vanishing cavity the values of this force are different. The difference between the values for two selected shapes of cavity is due to the area bounded internally by one cavity and externally by the other, or rather to the limit of this area as it vanishes. Hence we infer the important fact that the difference of the values of the force corresponding to different cavities is independent of the size and shape of the outer boundary of the surface distribution. It is simply a function of the shapes of the two cavities, proportional of course to the surface-density at P. Now since these values of a tangential component corresponding to different cavities have definite differences independent of the outer boundary of the distribution, their limits, as the outer boundary tends to any limiting form, must have the same definite differences; one may vanish, but certainly not all. [In the case of a plane surface of uniform density the one which vanishes would correspond to a cavity similar and similarly situated to the limiting form of the outer boundary, with P for centre of similitude.] Thus the result apparently obtained in § 47 could not be true.

49. At this stage it is natural to note the probability that the exclusion of infinitely great differences between the values of the force for different cavities implies a restriction on the nature of the contemplated cavities and their mode of vanishing. And we therefore consider for a moment the question of what kinds of passage to limit make the semi-convergent force-integrals converge to definite values.

An answer, though not a complete one, may be founded upon a well-known theorem already made use of without explicit quotation at the end of § 48. The theorem is that an annular plane lamina of uniform surface-density, whose inner and outer boundaries are similar and similarly situated curves, exerts no tangential attraction at the centre of similitude. Applying this result to the part of a uniform disc contained between two successive positions of a contour which is diminishing in size without change of shape round a fixed centre of similitude P, we find that the shrinkage of the contour makes no difference in the value of the tangential force at P due to the part of the disc outside the contour; accordingly the limit of this force is definite for vanishing of the cavity. If we have to do with a surface-density which is not uniform, residing in a curved surface, the errors involved in neglecting the variability of the surface-density and the curvature of the surface can be rendered as small as we please by taking a sufficiently small contour, provided the curvature of the surface is definite and the surface-density continuous at P. Hence the reasoning for a plane uniform surface can be rendered applicable to the more general case, and we arrive at the following result:—The semi-convergent integral representing tangential force at any point P of a surface distribution is rendered convergent by selecting a cavity of definite shape, and diminishing to zero, without change of shape, the scale of the geometrical configuration consisting of the cavity and the point P.

50. A wider range of cases of convergence might be obtained by study of the form of the integral representing the component of attraction in the direction of the axis of x, at an origin situated in a plane disc of uniform surface-density σ, occupying part of the plane $z=0$. The integral is, in the notation of polar coordinates,

$$\iint \sigma r^{-1} \cos\theta\, dr\, d\theta,$$

and if the external boundary and the boundary of the cavity are respectively $r=F(\theta)$ and $r=f(\theta)$, this reduces to

$$\sigma \int_0^{2\pi} \{\log F(\theta) - \log f(\theta)\} \cos\theta\, d\theta.$$

The danger of an infinity here arises out of the tendency of $f(\theta)$ to zero, as the inner boundary closes in round the origin. But if the form of $f(\theta)$ is such that we can write it $\eta\phi(\theta)$, where η is independent of θ and tends to zero with the vanishing of the cavity, while $\phi(\theta)$ is neither zero nor infinite for any value of θ, we may put

$$\log f(\theta) = \log \eta + \log \phi(\theta),$$

and the only term threatening an infinity is now

$$\sigma \log \eta \int_0^{2\pi} \cos\theta\, d\theta,$$

which vanishes for all values of η different from zero, and therefore has the limit zero for $\eta \to 0$.

If $\phi(\theta)$ is independent of η we have the case discussed otherwise in § 49. If $\phi(\theta)$ involves η, but in such a way that its order of magnitude is not determined by that of η, we have a type of convergence more general than that of § 49, but not of a seriously different character.

We note here that the disappearance of the term which threatened infinity is bound up with one special restriction on the cavity contemplated, namely that the contour of the cavity, i.e. the nearer edge of the matter, must completely surround the point at which the attraction is estimated, otherwise the integrals of $\cos\theta$ and $\sin\theta$ would not vanish. This explains the fact that at a point on the edge of a disc the attraction is infinite.

Another feature of the cavities now under discussion is that the distances from the origin of the various points of the edge of the cavity all become small of the same order as the cavity closes in.

51. The class of cavity just described (which it will be convenient to call class α) does not, of course, represent the only mode of making the force integrals converge to definite values. It is to be expected that there are other classes of vanishing cavity capable of producing convergence, and of these an important example is that which comprises such as are symmetrical about two axes at right angles through the point P. For the contours of such cavities, both

$$\int \log f(\theta) \cos\theta d\theta \quad \text{and} \quad \int \log f(\theta) \sin\theta d\theta$$

vanish, so that both components of tangential force are definite. In this case the value of each force-component is independent of the shape of the cavity, so that the vanishing of the cavity need not be effected by keeping the shape unaltered and gradually diminishing the scale; for it is permissible, while the linear dimensions are diminished, to keep changing the shape, so long as the symmetry conditions are never infringed.

For example, the cavity might be a rectangular slit whose length, though tending to zero, tends to be infinitely great in comparison with the breadth; here the distances of the various points of the edge of the cavity do not all become small of the same order.

The class of symmetrical cavities we may call class β.

52. In the present application, in order to be able to employ the foregoing principles of semi-convergence, we shall suppose that the surface-density in Ω is a continuous function at the points where Ω meets S, which includes the supposition that the material distribution in Ω does not terminate on the surface S, but passes through it. We postulate, further, that the force shall always be reckoned for a cavity of the class leading to finite values. And now we see that, since the differences of the values of the force for different cavities of this class are definite and independent of ϵ, one value of F' may tend to zero as $\epsilon \to 0$, but certainly not all; but, if one has a definite limit for $\epsilon \to 0$, then also all the others have definite limits.

53. Now let us return to the argument in § 47 by which the attempt was made to prove that a force-component G', specified by a cavity θ of assigned shape, tends to zero with ϵ. The reasoning broke down because the boundary of the portion whose attraction was denoted by G_2' ceased to be the curve θ. In the construction by which it was proposed to make $|G'|$ less than an arbitrarily assigned small quantity ω, the dimensions of θ were first selected, and *afterwards* a value of ϵ was chosen such that for it and all smaller values $|G_1'| < \frac{1}{2}\omega$. This order of choice of the dimensions of θ and ϵ implies that the effective contour of the portion whose attraction is G_2' is a quadrilateral on the surface Ω bounded by parallel curves on the surfaces S_1 and S_0 and by two opposite portions of the curve θ; and the breadth of the quadrilateral between the former curves may be taken as tending to infinite smallness in comparison with the length from one to the other of the opposite portions of the curve θ.

The limit form of this quadrilateral contour is simply the kind of rectangle quoted above as an example of a cavity of the class β. True it is not plane, not rectilineal, and not right-angled. But, given simple analytical circumstances, small errors may always be considered separately, and therefore when the curves and surfaces concerned are free from geometrical singularity the errors due to neglect of their curvatures may always be rendered as small as we please by making ϵ and the dimensions of θ sufficiently small. And when the quadrilateral may be as narrow as we please in comparison with its length, the contributions of its short and relatively distant ends to the contour integral may be made as small (relatively to the whole) as we please. Thus the possible obliquity of the ends is unimportant, and the contour may be regarded as tending ultimately to the narrow rectangular form.

Thus the reasoning which failed to prove that G' tends to zero with ϵ, when calculated for a cavity of the shape θ, does actually prove that G' tends to zero with ϵ when calculated for a particular cavity of the class β, and therefore also for all cavities of the class β.

Now some cavities of the class β belong also to the class α; therefore G' tends to zero with ϵ when calculated for some cavities of the class α. Therefore G' tends to a definite limit, as $\epsilon \to 0$, when calculated for any selected cavity of the class α.

54. This is just the result which we require in order to demonstrate the vanishing of $\operatorname*{Lim}_{\epsilon \to 0} \int N'dS$, when there is present a surface distribution of finite surface-density. For N' tends to zero at all points of S except those on a certain curve, the intersection of S and Ω; and non-zero values, provided they are definite, confined merely to a curve on the surface of integration, do not contribute to the value of the integral. Now we have shewn that, at points on the curve, N' calculated for any cavity of the class already discussed has a finite limit for $\epsilon \to 0$. Hence

$$\operatorname*{Lim}_{\epsilon \to 0} \int N'dS = 0.$$

55. The case in which the angle χ of intersection of the surfaces S and Ω vanishes at a point P is not covered by the preceding reasoning, and it is desirable to give a brief outline of the manner in which it may be discussed. In this case the surfaces Ω and S touch at the point P, and it is desired to prove that at P

$$\operatorname*{Lim}_{\epsilon \to 0} N' = 0.$$

The portion of Ω contained between S_1 and S_0 is not now a long narrow strip, and we must examine the shape of that part of its boundary which is nearest to P. Let us suppose (noting that the supposition excludes geometrical singularities in Ω and S) that with P as origin and suitably chosen axes the part of S near P is approximately

$$2z = ax^2 + 2hxy + by^2,$$

and the equation to the part of Ω near P is approximately

$$2z = a'x^2 + 2h'xy + b'y^2;$$

then the equations of the parts of S_1 and S_0 near to P are (to a sufficient approximation for the present purpose)

$$2z = \pm 2\epsilon + ax^2 + 2hxy + by^2.$$

Hence the projections on the tangent plane at P of the parts of the curves of intersection of Ω with S_1 and S_0 which are near to P are given approximately by the equations

$$\pm 2\epsilon = (a' - a)\, x^2 + 2\,(h' - h)\, xy + (b' - b)\, y^2,$$

and are, in fact, curves analogous to the indicatrix of Ω and no less definite in character. These may be called 'quasi-indicatrices.'

If the surface Ω does not cross the surface S at P, i.e. if

$$(a'-a)(b'-b)>(h'-h)^2,$$

one of the quasi-indicatrices is imaginary, the other a real ellipse with P as centre. This ellipse is an approximation to the boundary of the part of Ω contained between S_1 and S_0, and tends to the same limit form closing in round P for $\epsilon \to 0$. The convergence of N (absolute in this case) is a guarantee that the limit, for $\epsilon \to 0$, of the corresponding N' is zero.

If $(a'-a)(b'-b)<(h'-h)^2$, so that the surfaces S and Ω cross one another at P, the two quasi-indicatrices are conjugate hyperbolas with P for centre, their vertices tending to coincidence with P as $\epsilon \to 0$. These hyperbolas are an approximation to the part of the boundary near P of the part of Ω contained between S_1 and S_0. Now describe round P, in the surface Ω, a curve θ which (since N' is absolutely convergent at P) may be taken to be a circle. The diagram would be an oblique curvilinear cross inside a circle. N' may be split up into N_1' due to the part of Ω between S_1 and S_0 but outside θ, and N_2' due to the part of Ω between S_1 and S_0 and inside θ, namely the cross-shaped area. Having selected any small quantity ω we can, on account of the absolute convergence of N', choose θ so small that $N_2'<\frac{1}{2}\omega$; and then, since when θ is once chosen P is definitely outside the distribution that gives rise to N_1', we can choose ϵ so small that $N_1'<\frac{1}{2}\omega$; hence we have been able to choose ϵ so small that $N'<\omega$. Thus $\lim\limits_{\epsilon \to 0} N'=0$.

Therefore the contact of Ω and S at P does not disturb the value of

$$\lim_{\epsilon \to 0} \int N' dS.$$

56. Thus, both for distributions of finite volume-density and for those which include finite surface-density, the validity of the Gauss theorem has been established.

XI. Some Hydrodynamical Theorems.

57. Uniqueness Theorems. A well-known application of Green's Theorem (Article 18, formula 6) is to prove that the problem of finding a potential function ϕ for a given definite region, which shall satisfy certain conditions at the surface or surfaces which constitute the boundary of the region, cannot have two essentially different solutions.

The nature of the boundary conditions depends on the particular physical application which is contemplated. Thus if ϕ is to be an electrostatic potential it will be required to have a constant value over each isolated portion of the boundary and to be such that the surface integral of its normal gradient over each such isolated portion has a prescribed value. If ϕ is to be the hydrodynamical velocity potential of a liquid in motion the normal gradient of ϕ at each point of the boundary is prescribed, being required to be equal to the normal component of the velocity of the moving boundary.

In the simpler cases $\Delta\phi = 0$, and the proof is got by assuming two different ϕ's, ϕ_1 and ϕ_2, which satisfy the prescribed conditions, and by substituting $\phi_1 - \phi_2$ for V in formula (8) of Article 18. This gives

$$\int (\phi_1 - \phi_2)\left(\frac{\partial\phi_1}{\partial\nu} - \frac{\partial\phi_2}{\partial\nu}\right) dS = \int \Sigma \left\{\frac{\partial}{\partial x}(\phi_1 - \phi_2)\right\}^2 d\tau,$$

and the surface conditions are such as to ensure the vanishing of the surface integral; hence the volume integral must vanish, and as the subject of integration cannot be negative anywhere it must vanish everywhere in the region of integration.

Less simple cases are the electrical applications to conductors situated in a heterogeneous dielectric, or in a region in which there are interfaces between different dielectrics. Both the enunciations and the proofs of the theorems require care but present no serious difficulty.

58. As has been already suggested in Article 39, the application of a theorem of this type to regions which have no outer boundary and so 'extend to infinity' is not merely the taking of a particular case of a general theorem, but involves an additional step of passage to limit which requires special justification. No doubt the applied mathematician who is not disposed to give time to refinements of logic will take many such passages to limit on trust, without a qualm, for he has convictions based on physical considerations as weighty as any reasoning by pure mathematics. But, in approaching the usual hydrodynamical applications of such theorems, the more one thinks of an infinitely extended absolutely incompressible liquid, a system which instantaneously transmits force and energy to unlimited distance, the more one realises that it has no proper place in physics, and that (however useful it may be as an approximation to physical circumstances) it is a conception of the pure mathematician and must be studied by purely mathematical methods.

It is therefore within the scope of the present Tract to enquire whether for a region occupied by liquid, bounded internally by the surfaces of solids but without external boundary, there can be established any theorem corresponding to the fundamental 'uniqueness theorem' for hydrodynamical velocity potential, namely that, for a region bounded internally by closed surfaces which it surrounds and externally by a containing surface, two solutions of Laplace's equation having a prescribed normal gradient at each point of the complete boundary can differ only by a constant.

To take the known theorem and press it to a limit by endless extension of the containing boundary surface would be a task of some difficulty. For the functions dealt with are dependent in form upon the form of the boundary and must change as it changes, so that the volume integral and one of the surface integrals which appear in the proof would have not only changing regions of integration but also changing subjects of integration. It is best therefore to begin with a region which is externally unbounded, and to consider functions ϕ which satisfy the equation $\Delta\phi = 0$ at all points in this region.

59. Let us denote by dS an element of area of the closed surface or surfaces which constitute the inner boundary of such an infinite region, and by $d\nu$ an element of the outward-drawn normal at a point of such a surface. Let us also take another closed surface, whose element of area we may denote by $d\sigma$, which surrounds all the surfaces S. We begin by applying to the region which is bounded internally by S and externally by σ the theorem of Article 18, formula (8). If V is a function which satisfies the equation $\Delta V = 0$ at all points in the infinite region bounded internally by S, it does so at every point of the part of this region which lies within σ; hence the volume integral on the left-hand side of the equality vanishes, and we have

$$\int_\sigma V\frac{\partial V}{\partial \nu}\,d\sigma - \int_S V\frac{\partial V}{\partial \nu}\,dS = \int_S^\sigma \Sigma\left(\frac{\partial V}{\partial x}\right)^2 d\tau \quad \text{.........(29)}.$$

Now suppose that the surface σ expands without limit, so that the distance of every part of it from some definite origin O tends to infinite greatness; then clearly the volume integral and the first surface integral either both do or both do not converge to definite limit values, and if the former alternative obtains the two convergences are either both dependent on or both independent of the manner or form in which the surface σ tends to infinity.

On applying the theorem of Article 36 it appears that the volume integral converges absolutely, i.e. independently of the manner or form in which σ tends to infinity, provided $\Sigma\left(\frac{\partial V}{\partial x}\right)^2$, or (as we may call it for brevity) q^2, is such that for all values of the distance r measured from O greater than a definite length a

$$q^2 < Mr^{-\mu} \quad\text{..........................(30)},$$

where M is a constant and $\mu > 3$.

Under these circumstances the surface integral over σ also tends to a definite limit whose value may be calculated for any special form of σ which is convenient. If σ be taken to be a sphere with O as centre and r as radius, $d\sigma = r^2 d\omega$ where $d\omega$ is an element of solid angle, and $\partial V/\partial\nu$ is the same as $\partial V/\partial r$; as q is, for great values of r, of the order of greatness of $r^{-\frac{1}{2}\mu}$ at most, $\partial V/\partial r$ is at most of the same order of greatness, and V at most of the order $r^{-\frac{1}{2}\mu+1}$. Hence the surface integral is at most of the order of greatness of

$$\int r^{3-\mu} d\omega$$

and therefore tends to the limit zero for $r \to \infty$.

Thus we have, as the result of passage to limit, the theorem

$$-\int_S V \frac{\partial V}{\partial \nu} dS = \int_S^\infty \Sigma\left(\frac{\partial V}{\partial x}\right)^2 d\tau \quad\text{..............(31)},$$

valid for functions V which satisfy the above specified conditions.

60. In the hydrodynamical application, where V is a velocity potential, q is the resultant velocity, and the inequalities

$$q^2 < Mr^{-\mu}, \quad \mu > 3 \quad\text{......................(32)}$$

take the place of the common but only imperfectly intelligible statement that the velocity 'vanishes at infinity.' To the physicist, however, the interpretation of this restriction on V which appears most significant is that which is expressible in terms of the kinetic energy, namely that the motion is one having a definite amount of kinetic energy. If q^2 were of the order of r^{-3} or of a greater order of magnitude than r^{-3} the integral representing the total kinetic energy would almost certainly tend to indefinite greatness.

When the kinetic energy of the motion is definite formula (31) gives an expression of its value as a surface integral over the surface S.

61. It may be remarked here that the inequalities (32) allow of fractional values of μ provided only that $\mu > 3$. It is known however from the general theory of the solutions of Laplace's equation* that, in problems dealing with regions of the same general character as the region surrounding a closed surface S, fractional powers of r do not occur. So we may think of $\frac{1}{2}\mu - 1$, the negative power of r associated with V, as integral, and of μ as an even integer.

62. A uniqueness theorem for the infinite region under consideration is obtained as follows. Let ϕ_1 and ϕ_2 be two functions which satisfy Laplace's equation at all points of the region extending from S to infinity, which have equal normal gradients at all points of the surfaces S, and to each of which corresponds a liquid motion having a definite (i.e. finite but not prescribed) amount of kinetic energy. In equation (31) substitute $\phi_1 - \phi_2$ for V and we get

$$-\int (\phi_1 - \phi_2)\left(\frac{\partial \phi_1}{\partial \nu} - \frac{\partial \phi_2}{\partial \nu}\right) dS = \int \Sigma \left\{\frac{\partial}{\partial x}(\phi_1 - \phi_2)\right\}^2 d\tau.$$

Now the left-hand side vanishes because $\frac{\partial \phi_1}{\partial \nu} = \frac{\partial \phi_2}{\partial \nu}$ at S, and consequently the volume integral must vanish; on account of the positive character of the subject of integration this requires that at every point

$$\Sigma \left\{\frac{\partial}{\partial x}(\phi_1 - \phi_2)\right\}^2 = 0.$$

Hence ϕ_1 and ϕ_2 cannot differ except by a constant.

63. Theorems concerning Kinetic Energy†. It is well known that, for liquid in a given region whose boundaries are moving in a prescribed manner, the irrotational motion has less kinetic energy than any possible rotational motion. The following theorem, which is in a certain sense a particular case of the former, is believed to be new, and is given here because of the important dynamical principles which can be deduced from it.

In any region bounded by given surfaces moving in given manners consider alternatively two possible liquid motions, (α) continuous irrotational motion, or (β) several continuous irrotational motions in various sub-regions separated from one another by surfaces at which there is continuity of normal but not of tangential velocity. The kinetic energy

* Cf. Thomson and Tait, *Natural Philosophy*, Edition of 1890, Vol. I. p. 181.

† The results of Articles 63 to 65 were obtained independently by the author, and so far as he knows they have not been previously published. But on enquiry he has found that at least theorems (i) and (ii) of Article 64 have been familiar for some time to Dr Bromwich and Mr J. H. Grace.

of the motion (β) *is greater than that of the motion* (α) *by an amount equal to the kinetic energy of such motion as would have to be superposed on* (α) *in order to produce* (β).

The proof, which consists of a simple application of the theorems of Article 18, varies slightly in detail according to the nature of the region and the surfaces dealt with. Let us consider a region bounded internally by a surface S and externally by a containing surface σ, and let each of these surfaces be moving (not necessarily rigidly) with velocity whose normal component at any point is typified by V and v respectively. Let $d\nu$ represent an element of normal drawn outwards from any closed surface, and let the density of the liquid be taken as unity.

Consider (α) a continuous motion in the region between S and σ, having a velocity potential ϕ; (β) a motion having a discontinuity of tangential flow over a closed surface S' which does not surround S, the motion having a velocity potential $\phi + \chi$ inside S' and a velocity potential $\phi + \psi$ outside S', and the normal velocity at S' being typified by V'.

Then ϕ, χ and ψ all satisfy Laplace's equation, and in addition the following surface conditions are satisfied:—

At S, $\qquad \partial\phi/\partial\nu = V, \quad \partial\psi/\partial\nu = 0;$

at σ, $\qquad \partial\phi/\partial\nu = v, \quad \partial\psi/\partial\nu = 0;$

at S', $\qquad \dfrac{\partial\phi}{\partial\nu} + \dfrac{\partial\psi}{\partial\nu} = V' = \dfrac{\partial\phi}{\partial\nu} + \dfrac{\partial\chi}{\partial\nu}.$

Denoting the kinetic energies by T with appropriate suffix, we know from Article 18, equation (8), that

$$T_\alpha = -\tfrac{1}{2}\int \phi \frac{\partial\phi}{\partial\nu} dS + \tfrac{1}{2}\int \phi \frac{\partial\phi}{\partial\nu} d\sigma,$$

$$T_\beta = -\tfrac{1}{2}\int (\phi + \psi)\left(\frac{\partial\phi}{\partial\nu} + \frac{\partial\psi}{\partial\nu}\right) dS$$
$$+ \tfrac{1}{2}\int (\phi + \psi)\left(\frac{\partial\phi}{\partial\nu} + \frac{\partial\psi}{\partial\nu}\right) d\sigma - \tfrac{1}{2}\int (\phi + \psi)\left(\frac{\partial\phi}{\partial\nu} + \frac{\partial\psi}{\partial\nu}\right) dS'$$
$$+ \tfrac{1}{2}\int (\phi + \chi)\left(\frac{\partial\phi}{\partial\nu} + \frac{\partial\chi}{\partial\nu}\right) dS'.$$

Hence, remembering the surface conditions, we get

$$T_\beta - T_\alpha = -\tfrac{1}{2}\int \psi \frac{\partial\phi}{\partial\nu} dS + \tfrac{1}{2}\int \psi \frac{\partial\phi}{\partial\nu} d\sigma$$
$$+ \tfrac{1}{2}\int (\chi - \psi) \frac{\partial\phi}{\partial\nu} dS' + \tfrac{1}{2}\int \chi \frac{\partial\chi}{\partial\nu} dS' - \tfrac{1}{2}\int \psi \frac{\partial\psi}{\partial\nu} dS'.$$

But by Green's Theorem (Article 18, formula 6), applied to the region outside S'

$$-\int \psi \frac{\partial \phi}{\partial \nu} dS + \int \psi \frac{\partial \phi}{\partial \nu} d\sigma - \int \psi \frac{\partial \phi}{\partial \nu} dS'$$

$$= -\int \phi \frac{\partial \psi}{\partial \nu} dS',$$

and by the same theorem applied to the region inside S'

$$\int \chi \frac{\partial \phi}{\partial \nu} dS' = \int \phi \frac{\partial \chi}{\partial \nu} dS' = \int \phi \frac{\partial \psi}{\partial \nu} dS',$$

so that

$$T_\beta - T_\alpha = \tfrac{1}{2}\int \chi \frac{\partial \chi}{\partial \nu} dS' - \tfrac{1}{2}\int \psi \frac{\partial \psi}{\partial \nu} dS'$$

$$= \tfrac{1}{2}\int^{S'} \Sigma \left(\frac{\partial \chi}{\partial x}\right)^2 d\tau + \tfrac{1}{2}\int_{SS'}^{\sigma} \Sigma \left(\frac{\partial \psi}{\partial x}\right)^2 d\tau \quad \ldots\ldots\ldots\ldots(33);$$

the final expression is essentially positive, being the kinetic energy of the motion (represented by χ and ψ) which if superposed on (α) would yield (β).

The internal boundary S is not necessary, but gives generality to the theorem; clearly it may consist of one or of several distinct closed surfaces. S' also may be regarded as typical of several closed surfaces exterior to one another, or some surrounding others. The theorem also holds good if S' cuts any of the surfaces S and σ, being bounded by the curves of section, or if S' surrounds some or all of the surfaces S; for such cases the proof would require slight and fairly obvious modifications which need not be set out in detail. Enough has been said to establish the truth of the theorem as enunciated.

64. Let us now pass to the consideration of some special cases of the general theorem.

(i) Suppose S' to surround S but to be wholly inside σ, and let both S' and σ be at rest; and let the motion of S be any motion which is compatible with rigidity of each of the surfaces S. Then in the β motion the complete boundary of the region between S' and σ is at rest, and so there is no motion there, i.e. $\chi = -\phi$. We may think of S as made up of the surfaces of solid bodies moving in the liquid, and S' in the β motion as a new fixed outer boundary substituted for the fixed outer boundary σ of the α motion. Hence the theorem: *Any number of solid bodies are moving with given linear and angular velocities in homogeneous liquid which is bounded by a fixed outer boundary. If for this outer boundary there were substituted another fixed boundary lying completely inside the former one, the kinetic energy of the liquid motion would be increased by an amount equal to the kinetic energy of the motion which would have to be superposed on the former motion in order to produce the latter.*

Of course the outer boundary of the original motion may be at infinity, provided the motion has definite kinetic energy. In this case the new boundary S' might be a plane or any open surface extending to infinity.

The solid bodies and the liquid constitute a dynamical system whose motion is determined by the motion of the solids, so that it has six times as many coordinates as there are movable solids. The inertia coefficients depend on the configuration, including the shape and position of the boundary. An increase of kinetic energy for given velocities of the solids means an increase of the inertia coefficients. Hence our theorem tells us that a closing in of the fixed boundary involves increase of inertia.

Thus it might be expected, for example, that a submarine vessel would be more difficult to propel or to steer when near to the bottom of the sea, or to the shore, than when out in the open deep sea.

(ii) As a second special case suppose S' to be wholly inside σ but not to surround S, and let both S' and σ be at rest, while S moves in any manner compatible with the rigidity of each of the surfaces typified by S. As before we think of S and S' as rigid material boundaries, and note that in the β motion there is no motion inside S'. Hence the theorem: *Any number of solid bodies are moving with given linear and angular velocities in homogeneous liquid having a fixed outer boundary. If another fixed solid were present the kinetic energy of the liquid motion would be greater than it actually is by an amount equal to the kinetic energy of the motion which would have to be superposed on the first motion in order to produce the second.*

This theorem indicates that the effect of the presence of a fixed solid is to increase the effective inertia coefficients of movable solids in its neighbourhood.

(iii) As a third special case suppose S' to be wholly inside σ but not to surround S, let σ be at rest, and let both S' and S be moving in any manner compatible with the rigidity of each separate surface. As before we think of S as made up of the surfaces of moving solid bodies, and in the first instance we think of S' as a rigid massless material shell. There is then, in the β system, motion inside as well as outside S', and we may conveniently split T_β into two parts T_β' for the motion outside S' and T_β'' for the motion inside S'. The general theorem now takes the form

$$T_\beta' + T_\beta'' - T_\alpha = T(\psi) + T(\chi) \quad \text{..............(34)},$$

the meaning of the symbols on the right-hand side being obvious.

Now suppose the liquid inside S' to be replaced by solid matter, whose kinetic energy in the given motion of S' is t. If $t \geqslant T_\beta''$ our equality leads to the inequality

$$T_\beta' + t > T_\alpha + T(\psi) + T(\chi) > T_\alpha \quad \text{..............(35)}.$$

Hence the theorem: *Any number of solid bodies are moving with given linear and angular velocities in homogeneous liquid having a fixed outer boundary. If in addition there were present another solid body moving in any manner the kinetic energy of the motion of the liquid and the new solid would be together greater than the kinetic energy of the original fluid motion, provided the new solid has for its given motion not less kinetic energy than that of the irrotational motion of liquid occupying a boundary similar to the boundary of the solid and moving in a similar manner.*

It may be remarked that the motion of a liquid as if solid, when not a motion of mere translation, is rotational, and so has greater kinetic energy than the irrotational motion having the same boundary. Hence the condition $t \geqslant T_\beta'$ is certainly satisfied if the solid body S' is homogeneous and of the same specific gravity as the liquid. And as the moving of matter to the boundary of a solid, without change of total mass, increases the moments of inertia, a hollow solid having the same mass as the liquid it displaces would have not less kinetic energy. Hence the condition $t \geqslant T_\beta'$ is likely to be satisfied for many solids which are not lighter than the liquid they displace.

This theorem accordingly indicates that the effect of the presence of a movable solid which is not lighter than the liquid it displaces is generally to increase the kinetic energy of the total motion, and therefore to increase the effective inertia coefficients of movable solid bodies in its neighbourhood.

It is readily seen that, when the boundary of the additional solid body is given, its total mass and the distribution of its mass within its boundary can affect only those coefficients which multiply the squares and products of the fluxes of the six coordinates of the body itself in the expression for the total kinetic energy. All the other coefficients may be affected by the geometrical boundary-configuration but not by the mass-configuration of the new solid. Hence the theorem, as regards the inertia coefficients of the original solids, is perfectly general.

65. 'Suction.' Let T be the kinetic energy and U the work function of the acting forces for the dynamical system consisting of one or more solid bodies moving in liquid, and let θ be typical of the generalised coordinates of the system. In the Lagrangian equation of motion

$$\frac{d}{dt}\left(\frac{\partial T}{\partial \dot{\theta}}\right) = \frac{\partial T}{\partial \theta} + \frac{\partial U}{\partial \theta}$$

the term $\partial T/\partial\theta$ represents an inertia effect which can in a certain sense be regarded as equivalent to a force tending so to modify the configuration as to increase T, just as $\partial U/\partial\theta$ is a force tending so to modify the configuration as to increase U. Thus when the kinetic energy of a system is a function not only of the time-fluxes of the coordinates but also of the coordinates themselves there are apparent forces, which are really inertia effects, making for increase of the kinetic energy.

Now from the three dynamical theorems stated above, and from others on similar lines which it would be easy to formulate, it is fairly clear that generally the approach of a movable solid to a fixed boundary or to another solid which is held fixed, or even to another movable solid, so changes the configuration as to increase the kinetic energy. In some cases this is capable of complete logical proof, as for example when a single solid moves in liquid which has no boundary except a single infinite plane. In other cases it is difficult to distinguish exhaustively between changes of configuration which tend to increase the kinetic energy and those which tend to decrease it, but various kinds of change can be assigned to one class or the other with such a high degree of probability as is equivalent to certainty for practical purposes. Generally the question at issue is whether what has been called $T(\psi)$ is increased or decreased by the contemplated change of configuration, and one feels justified in stating (though the term used is not precise) that $T(\psi)$ increases with the proximity of two bodies, or of one body and a fixed boundary.

Hence the inertia term $\partial T/\partial\theta$ usually manifests itself as a force making for increase of proximity, as it were an attraction between the bodies or the body and the boundary. This is what is called 'suction.' It is an additional effect to the increase of inertia previously discussed. If, for example, a submarine were passing near another vessel the theory points not only to abnormal heaviness in steering and propelling, but also to the possibility of the steering being utterly vitiated by forces and couples due to suction.

66. Semi-convergent Volume Integrals to Infinity. The theorem of Article 36 suggests the rough rule that a volume integral to infinity whose subject of integration f at great distance r from a definite origin O tends to smallness of the order $r^{-\mu}$ is convergent if $\mu > 3$, semi-convergent or divergent if $\mu = 3$, divergent if $\mu < 3$. This is, however, by no means an accurate statement, for the divergence theorem analogous to that of Article 14 is as follows: *If at all points outside a sphere, having O as centre and a definite radius a, f is algebraically greater than $mr^{-\mu}$, where m is a constant greater than zero and $\mu \leqslant 3$, the integral $\int f d\tau$, taken through a region whose outer boundary tends to infinite remoteness from O in all directions, is divergent.* This theorem indicates that for $\mu \leqslant 3$ there is no possibility of convergence if f is a function which has (outside the sphere a) everywhere the same sign and is such that $r^{\mu} f$ is everywhere definitely different from zero, but it by no means shuts the door on semi-convergence, i.e. convergence associated with some special mode of infinite widening of the outer boundary, if f changes sign from place to place.

67. A criterion for the existence of special modes leading to convergence, and a complete specification of them (when they exist) for a general subject of integration, would probably be extremely difficult to obtain. But there are two particular types of subject of integration for which certain modes leading to convergence can be readily recognised.

(i) Using spherical polar coordinates r, θ, ϕ, let us first suppose f to be of the form

$$r^{-3}\psi(\theta, \phi) + g,$$

where ψ is a finite single-valued function of angular position which satisfies the condition

$$\int_0^{\pi} d\theta \int_0^{2\pi} d\phi \sin\theta\, \psi(\theta, \phi) = 0,$$

and g is a function of position which tends to smallness of a higher order than r^{-3}. The volume integral of g in general converges absolutely. As regards the first term of f, let us consider its integral through the volume contained between the two similar and similarly situated boundaries, having the origin for centre of similitude, whose equations are

$$r = \alpha F(\theta, \phi),$$
$$r = \beta F(\theta, \phi),$$

where F is a function which is always definitely greater than zero, and α and β are positive parameters of which α is the greater. The volume integral is

$$\iiint r^{-3}\psi(\theta, \phi)\, r^2 \sin\theta\, d\theta\, d\phi\, dr,$$

which becomes, on integration with respect to r,

$$\log(\alpha/\beta) \iint \psi(\theta, \phi) \sin\theta\, d\theta\, d\phi,$$

which is zero by hypothesis.

Passing now to the volume integral of $r^{-3}\psi(\theta, \phi)$ through a region whose outer boundary is $r = \beta F(\theta, \phi)$, we see from the above that the value is the same as if we had integrated through the wider region whose outer boundary is $r = \alpha F(\theta, \phi)$, no matter how great α may be. Thus the integral out to the α surface has a definite constant value, and therefore a definite limit value, while α becomes great without limit; the value of course depends on the form of the function $F(\theta, \phi)$. In other words our semi-convergent integral is rendered convergent by selecting an outer boundary of arbitrary but definite shape, and a definite origin O within it, and by increasing indefinitely without change of shape the scale of the geometrical configuration consisting of the outer boundary and the point O.

It will be noticed that the property of $\psi(\theta, \phi)$ which leads to this convergence is a property of Laplace's functions of integral order other than zero. Hence ψ may be the sum of any number of such complete surface harmonics*. A particular case of importance is that in which f is a finite single-valued solution of Laplace's equation, for then ψ is a complete surface harmonic of order 2.

68. The proof suggests a certain slight and perhaps unimportant extension of the theorem. Instead of $r = \alpha F(\theta, \phi)$ we might take the outer boundary to be

$$r = \alpha^{\chi(\theta, \phi)} F(\theta, \phi),$$

where χ is a function which is positive (so that the outer and inner boundaries may not intersect) and free from infinities; the volume integral between this and the surface $r = F(\theta, \phi)$ is

$$\log \alpha \iint \chi(\theta, \phi)\, \psi(\theta, \phi) \sin\theta\, d\theta\, d\phi.$$

* Any single-valued function of angular position, provided it be of limited variation, can be expanded in a uniformly convergent series of Laplace's functions. (Jordan, *Cours d'Analyse*, t. II, § 244.) If the Laplace's function of order zero is absent from the expansion of ψ, the condition for convergence of the volume integral is satisfied.

If χ and ψ be such as to make this vanish there is convergence for the mode of expansion of the outer boundary corresponding to $a \to \infty$. For example χ might be a constant plus a sum of surface harmonics of integral orders different from any which occur in ψ. It may be noted that χ may involve a without invalidating the argument*.

69. (ii) In the second place let us suppose f to be of the form

$$\lambda(r)\,\psi(\theta, \phi)$$

where ψ satisfies the same criterion as in Article 67, and $\lambda(r)$ is any function of r which does not become infinite for any definite value of r which occurs in any contemplated region of integration. Consider the volume integral of this f for the volume between the concentric spheres $r = a$ and $r = \beta$, where $a > \beta$. The integral is

$$\iiint \lambda(r)\,\psi(\theta, \phi)\, r^2 \sin\theta\, d\theta\, d\phi\, dr,$$

which becomes, on integration with respect to r,

$$\int_\beta^a r^2 \lambda(r)\, dr \iint \psi(\theta, \phi) \sin\theta\, d\theta\, d\phi,$$

which is zero in virtue of the hypothesis with regard to ψ.

From this it follows, by reasoning similar to that employed above, that the volume integral for this type of f is rendered convergent by taking as outer boundary a sphere whose centre is O and by increasing the radius of the sphere without limit.

Probably both this theorem and the preceding one could be generalised by taking other curvilinear coordinates instead of r, θ, ϕ.

70. It is clear that theorems analogous to those just established hold for the semi-convergence of certain integrals of the kind discussed in Section III, it being a question of the mode of closing in of a cavity instead of the mode of expansion of an outer boundary.

71. *The Integral of Linear Momentum in Hydrodynamics.* A familiar example of a semi-convergent volume integral to infinity is the

* It may be possible, even in cases where ψ does not comply with the hypothesis of Article 67, to secure convergence by giving a suitable form to χ. But in such cases the expansions of both ψ and χ contain the Laplace's function of zero order (i.e. a constant), and therefore the surface integral of their product over the unit-sphere contains at least one non-vanishing term. The vanishing of the whole is secured by providing for one or more other non-vanishing terms, each resulting from the integration of the product of two surface harmonics of the same order, with constants adjusted to give a zero sum, if this can be done without sacrificing the positive character of χ.

integral representing the total linear momentum, resolved in any direction, of the motion of unbounded liquid due to the motion of a solid body through it. When the velocity potential ϕ tends to smallness of the order r^{-2} a velocity component u generally tends to smallness of the order r^{-3}. And since ϕ satisfies Laplace's equation so also does $\partial\phi/\partial x$ or u. Hence the momentum integral $\int u\,d\tau$ is semi-convergent and converges if the outer boundary of the region of integration tends to infinity in any of the manners specified in Articles 67, 68 and 69.

72. *The Integral of Angular Momentum in Hydrodynamics.* In the integrals of the type

$$\int\left(y\frac{\partial\phi}{\partial z}-z\frac{\partial\phi}{\partial y}\right)d\tau,$$

which represent components of the moment of momentum of a liquid motion, when ϕ is of the order r^{-2} the subjects of integration are generally also of the order r^{-2}. But since ϕ satisfies Laplace's equation so also does $y\partial\phi/\partial z - z\partial\phi/\partial y$, and likewise the two similar expressions. Hence the volume integrals are of the class whose subjects of integration are finite single-valued solutions of Laplace's equation. The expansion of the subject of integration in a series of solid spherical harmonics gives a series of integrals of the type discussed in Article 69, which can be rendered convergent by always taking for outer boundary a sphere whose centre is the origin. Thus the angular momentum integrals are semi-convergent.

Printed in the United States
By Bookmasters